Recent Developments in the Theory of Lorentz Spaces and Weighted Inequalities

of the
American Mathematical Society

Number 877

Recent Developments in the Theory of Lorentz Spaces and Weighted Inequalities

María J. Carro
José A. Raposo
Javier Soria

May 2007 • Volume 187 • Number 877 (second of four numbers) • ISSN 0065-9266

American Mathematical Society
Providence, Rhode Island

2000 *Mathematics Subject Classification.*
Primary 42B25; Secondary 26D10, 46E30, 47B38, 47G10.

Library of Congress Cataloging-in-Publication Data

Carro, María J., 1961–
  Recent developments in the theory of Lorentz spaces and weighted inequalities / María J. Carro, José A. Raposo, Javier Soria.
    p. cm. — (Memoirs of the American Mathematical Society, ISSN 0065-9266 ; no. 877)
  "May 2007, volume 187, number 877 (second of 4 numbers)."
  Includes bibliographical references and index.
  ISBN 978-0-8218-4237-9 (alk. paper)
  1. Littlewood-Paley theory. 2. Lorentz spaces. 3. Inequalities (Mathematics) I. Raposo, José A. II. Soria, Javier, 1962– III. Title.
QA403.5.C367 2007
515'.2433–dc22   2007060666

Memoirs of the American Mathematical Society

This journal is devoted entirely to research in pure and applied mathematics.

Subscription information. The 2007 subscription begins with volume 185 and consists of six mailings, each containing one or more numbers. Subscription prices for 2007 are US$649 list, US$519 institutional member. A late charge of 10% of the subscription price will be imposed on orders received from nonmembers after January 1 of the subscription year. Subscribers outside the United States and India must pay a postage surcharge of US$38; subscribers in India must pay a postage surcharge of US$43. Expedited delivery to destinations in North America US$53; elsewhere US$130. Each number may be ordered separately; *please specify number* when ordering an individual number. For prices and titles of recently released numbers, see the New Publications sections of the *Notices of the American Mathematical Society*.

Back number information. For back issues see the *AMS Catalog of Publications*.

Subscriptions and orders should be addressed to the American Mathematical Society, P. O. Box 845904, Boston, MA 02284-5904, USA. *All orders must be accompanied by payment.* Other correspondence should be addressed to 201 Charles Street, Providence, RI 02904-2294, USA.

Copying and reprinting. Individual readers of this publication, and nonprofit libraries acting for them, are permitted to make fair use of the material, such as to copy a chapter for use in teaching or research. Permission is granted to quote brief passages from this publication in reviews, provided the customary acknowledgment of the source is given.

Republication, systematic copying, or multiple reproduction of any material in this publication is permitted only under license from the American Mathematical Society. Requests for such permission should be addressed to the Acquisitions Department, American Mathematical Society, 201 Charles Street, Providence, Rhode Island 02904-2294, USA. Requests can also be made by e-mail to reprint-permission@ams.org.

Memoirs of the American Mathematical Society is published bimonthly (each volume consisting usually of more than one number) by the American Mathematical Society at 201 Charles Street, Providence, RI 02904-2294, USA. Periodicals postage paid at Providence, RI. Postmaster: Send address changes to Memoirs, American Mathematical Society, 201 Charles Street, Providence, RI 02904-2294, USA.

© 2007 by the American Mathematical Society. All rights reserved.
Copyright of individual articles may revert to the public domain 28 years
after publication. Contact the AMS for copyright status of individual articles.
This publication is indexed in *Science Citation Index*®, *SciSearch*®, *Research Alert*®,
CompuMath Citation Index®, *Current Contents*®/*Physical, Chemical & Earth Sciences*.
Printed in the United States of America.

∞ The paper used in this book is acid-free and falls within the guidelines
established to ensure permanence and durability.
Visit the AMS home page at http://www.ams.org/

10 9 8 7 6 5 4 3 2 1   12 11 10 09 08 07

Per a la Beatriu

Contents

Abstract	ix
Foreword	xi
Introduction	1

1 Boundedness of operators on characteristic functions and the Hardy operator — 5
  1.1 Introduction 5
  1.2 Boundedness on characteristic functions 6
  1.3 The Hardy operator and the classes B_p 15

2 Lorentz spaces — 25
  2.1 Introduction 25
  2.2 $\Lambda_X^p(w)$ spaces 26
  2.3 Quasi-normed Lorentz spaces 34
    2.3.1 Absolutely continuous norm 35
    2.3.2 Density of the simple functions and L^∞ 40
  2.4 Duality 43
  2.5 Normability 73
  2.6 Interpolation of operators 80

3 The Hardy-Littlewood maximal operator in weighted Lorentz spaces — 89
  3.1 Introduction 89
  3.2 Some general results 91
  3.3 Strong-type boundedness in the diagonal case 97
  3.4 Weak-type inequality 105

3.5　Applications 120

Bibliography **123**

Index **127**

Abstract

The main objective of this work is to bring together two well known and, a priori, unrelated theories dealing with weighted inequalities for the Hardy-Littlewood maximal operator M. For this, we consider the boundedness of M in the weighted Lorentz space $\Lambda_u^p(w)$. Two examples are historically relevant as a motivation: If $w=1$, this corresponds to the study of the boundedness of M on $L^p(u)$, which was characterized by B. Muckenhoupt ([Mu]) in 1972, and the solution is given by the so called A_p weights. The second case is when we take $u=1$. This is a more recent theory, and was completely solved by M.A. Ariño and B. Muckenhoupt (see [AM2]) in 1991. It turns out that the boundedness of M on $\Lambda^p(w)$ can be seen to be equivalent to the boundedness of the Hardy operator A restricted to decreasing functions of $L^p(w)$, since the nonincreasing rearrangement of Mf is pointwise equivalent to Af^*. The class of weights satisfying this boundedness is known as B_p.

Even though the A_p and B_p classes enjoy some similar features, they come from very different theories, and so are the techniques used on each case: Calderón–Zygmund decompositions and covering lemmas for A_p, rearrangement invariant properties and positive integral operators for B_p.

It is our aim to give a unified version of these two theories. Contrary to what one could expect, the solution is not given in terms of the limiting cases above considered (i.e., $u=1$ and $w=1$), but in a rather more complicated condition, which reflects the difficulty of estimating the distribution function of the Hardy-Littlewood maximal operator with respect to general measures.

---

[1]Received by the editor October 10, 2000, and in revised form July 14, 2004.
2000 Mathematics Subject Classifications. Primary 42B25; Secondary 26D10, 26D15, 46E30, 47B38, 47G10.
Key words and phrases: Hardy operator, Hardy–Littlewood maximal function, Interpolation, Lorentz spaces, Weighted inequalities.
This work has been partially supported by grants MTM2004-02299 and 2005SGR00556.

Foreword

On September 4, 1998, José Antonio Raposo, the second named author, died when he was only 40. He had just received his Ph.D. degree a few months earlier, under our supervision. He was really happy for all the new projects he had for the future, and so were we, since he was an extraordinary mathematician, and a very valuable friend.

This work is an updated version of his thesis, written as a self-contained text, with most of the motivations, examples and applications available in the literature.

We want to thank all the people who have encouraged us to write this book, and specially José Antonio's family. We also thank Joan Cerdà who has read the whole manuscript and has given us many good advises, improving the final version of these notes.

María J. Carro and Javier Soria
Barcelona, September 2000

Introduction

The main objective of this work is to bring together two well known and, a priori, unrelated theories dealing with weighted inequalities for the Hardy-Littlewood maximal operator

$$Mf(x) = \sup_{x \in Q} \frac{1}{|Q|} \int_Q |f(y)|\, dy, \quad x \in \mathbb{R}^n.$$

For this, we consider the boundedness of M in the weighted Lorentz space $\Lambda_u^p(w)$, introduced by G.G. Lorentz ([Lo2]) in 1951,

$$M : \Lambda_u^p(w) \longrightarrow \Lambda_u^p(w). \tag{1}$$

Two examples are historically relevant as a motivation: If $w = 1$, (1) corresponds to the study of the boundedness

$$M : L^p(u) \longrightarrow L^p(u),$$

which was characterized by B. Muckenhoupt ([Mu]) in 1972, and the solution is given by the so called A_p weights. The second case is when we take in (1) $u = 1$. This is a more recent theory, and was completely solved by M.A. Ariño and B. Muckenhoupt (see [AM2]) in 1991. It turns out that the boundedness

$$M : \Lambda^p(w) \longrightarrow \Lambda^p(w),$$

can be seen to be equivalent to the boundedness of the Hardy operator A restricted to decreasing functions of $L^p(w)$, since the nonincreasing rearrangement of Mf is pointwise equivalent to Af^*. The class of weights satisfying this boundedness is known as B_p. Another results related to this problem can be found in [Sa], [Ne1], [CS1], [CS2], [Stp1], [Stp2], and [HM].

Even though the A_p and B_p classes enjoy some similar features, they come from very different theories, and so are the techniques used on each case: Calderón–Zygmund decompositions and covering lemmas for A_p, rearrangement invariant properties and positive integral operators for B_p.

It is our aim to give a unified version of these two theories. Contrary to what one could expect, the solution is not given in terms of the limiting cases above considered (i.e., $u = 1$ and $w = 1$), but in a rather more complicated condition, which reflects the difficulty of estimating the distribution function of the Hardy-Littlewood maximal operator with respect to general measures (some previous results on this direction can be found in [CHK] and [HK], for the particular case of the $L^{p,q}(u)$ Lorentz spaces with a power weight w).

In order to carry out this program as a self-contained monograph, we study in the first two chapters the main results needed for the rest of the book. Hence, in Chapter 1 we consider the boundedness of integral operators on monotone functions. The most important result is Theorem 1.2.11, where we prove that under very general conditions, the boundedness is determined by the action of the operator on some characteristic functions (see also [BPS] and [CS2]). We then consider the main example in this context, namely the Hardy operator, and introduce the B_p class.

In Chapter 2 we make an exhaustive study of the functional properties of the Lorentz spaces over general measure spaces, and arbitrary weights. In particular we consider the discrete case (sequence Lorentz spaces) where we answer several open questions about normability and duality (see Theorem 2.4.16). Some previous results were already proved (in the case w decreasing) in [Al] and [AEP].

Finally, in Chapter 3 we consider the solution to (1), which is proved in Theorem 3.3.5 (some sufficient conditions had been already obtained in [CS3] and [Ne2]). We see that the characterization that we obtain is based upon the so called $p - \epsilon$ condition, which is natural since both A_p and B_p enjoy this property. However is worth noticing that there exist non-doubling weights u for which M satisfies (1) for suitable weights w (see Theorem 3.3.10). We also study the weak-type and the restricted weak-type version of (1).

As far as possible, we have always tried to give precise bibliographic information about the results which were previously known, and also about the techniques used in the proofs, although due to the big number of people working in these theories, this a difficult task. We have also followed the standard notation found in the main reference books (e.g., [BS], [BL], [GR], [Ru], [Stn], [SW]).

Chapter 1

Boundedness of operators on characteristic functions and the Hardy operator

1.1 Introduction

This chapter includes general results on boundedness of operators on L^p such as the Hardy operator

$$Af(t) = \frac{1}{t}\int_0^t f(s)\,ds, \quad t > 0, \tag{1.1}$$

and the Hardy-Littlewood maximal operator

$$Mf(x) = \sup_{x \in Q} \frac{1}{|Q|}\int_Q |f(y)|\,dy, \quad x \in \mathbb{R}^n. \tag{1.2}$$

The main result is Theorem 1.2.11 and its Corollaries 1.2.12 and 1.2.14 where it is proved that a great variety of operators satisfy that their boundedness is characterized by the restriction to characteristic functions.

In the third section, we study the Hardy operator, recalling some known results that we shall use in the following chapters and including a detailed study of the discrete Hardy operator,

$$A_d f(n) = \frac{1}{n+1}\sum_{k=0}^{n} f(k), \quad n \in \mathbb{N}. \tag{1.3}$$

The following standard notations will be used very often: Letters such as X or Y are always σ-finite measure spaces and $\mathcal{M}(X)$ ($\mathcal{M}^+(X)$) denotes the space of measurable (resp. measurable and nonnegative) functions on X. The distribution function of f is

$$\lambda_f(t) = \mu(\{x : |f(x)| > t\}),$$

the nonincreasing rearrangement is

$$f^*(t) = \inf\{s > 0 : \lambda_f(s) \leq t\},$$

and

$$f^{**}(t) = \frac{1}{t}\int_0^t f^*(s)\,ds.$$

We denote by $\mathbb{R}^+ = (0,\infty)$. The letters w, $\tilde{w}$, w_0, ..., are used for weight functions in $\mathbb{R}^+$ (nonnegative locally integrable functions in $\mathbb{R}^+$). For a given weight w we write $W(r) = \int_0^r w(t)\,dt < \infty$, $0 \leq r < \infty$. If f is a positive nonincreasing (nondecreasing) function we will write $f \downarrow$ (resp. $f \uparrow$). The space L^p_{dec} (L^p_{inc}) is the cone of all nonincreasing (resp. nondecreasing) functions in L^p. Two positive quantities A and B, are said to be equivalent ($A \approx B$) if there exists a constant $C > 1$ (independent of the essential parameters defining A and B) such that $C^{-1}A \leq B \leq CA$. If only $B \leq CA$, we write $B \underset{\sim}{<} A$. The undetermined cases $0 \cdot \infty$, $\frac{\infty}{\infty}$, $\frac{0}{0}$, will always be taken equal to 0.

1.2 Boundedness on characteristic functions

In this section we consider the problem of characterizing the boundedness of

$$T : (L^{p_0}(X) \cap L, \|\cdot\|_{p_0}) \longrightarrow L^{p_1}(Y), \tag{1.4}$$

where L is a subclass of $\mathcal{M}(X)$ and T is an operator (usually linear or sublinear). That is, we are interested in studying the inequality

$$\|Tf\|_{L^{p_1}(Y)} \leq C\|f\|_{L^{p_0}(X)}, \quad f \in L. \tag{1.5}$$

In some cases inequality (1.5) holds for every $f \in L$ if and only if it holds on characteristic functions $f = \chi_A \in L$. This happens (see [CS2] and [Stp1]) in the case $L = L^{p_0}_{\text{dec}}(w_0)$, $X = (\mathbb{R}^+, w_0(t)\,dt)$ and $Y = (\mathbb{R}^+, w_1(t)\,dt)$ in the range of indices $0 < p_0 \leq 1$, $p_0 \leq p_1 < \infty$ and operators of the type

$$Tf(r) = \int_0^\infty k(r,t)f(t)\,dt, \quad r > 0, \qquad (1.6)$$

for positive kernels k. Using this, one can easily obtain a useful characterization of (1.4). For example, in the previous case the condition on the weights w_0, w_1 (nonnegative locally integrable functions in $\mathbb{R}^+$) for which $T: L^{p_0}_{\text{dec}}(w_0) \to L^{p_1}(w_1)$ (defined by (1.6)) is:

$$\left(\int_0^\infty \left(\int_0^r k(s,t)\,dt\right)^{p_1} w_1(s)\,ds\right)^{1/p_1} \leq C \left(\int_0^r w_0(s)\,ds\right)^{1/p_0}, \quad r > 0.$$

The range $0 < p_0 \leq 1$, $p_0 \leq p_1 < \infty$ is fundamental and, in fact, the previous principle also works in other situations, as in the case of increasing functions (see [Stp1])

$$T: L^{p_0}_{\text{inc}}(w_0) \to L^{p_1}(w_1)$$

where T is as in (1.6).

As we shall see (Theorem 1.2.11) all this is a consequence of a general principle that can be applied to a more general class of operators than those of integral type and for a bigger class of functions that includes the monotone functions.

We first need some definitions.

Definition 1.2.1 We say that $\emptyset \neq L \subset \mathcal{M}(X)$ is a regular class in X if, for every $f \in L$,

(i) $|\alpha f| \in L$, for every $\alpha \in \mathbb{R}$,

(ii) $\chi_{\{|f|>t\}} \in L$, for every $t > 0$, and

(iii) there exists a sequence of simple functions $(f_n)_n \subset L$ such that $0 \leq f_n(x) \leq f_{n+1}(x) \to |f(x)|$ a.e. $x \in X$.

Example 1.2.2 (i) If X is an arbitrary measure space, every functional lattice in X (i.e., a vector space $L \subset \mathcal{M}(X)$ with $g \in L$ if $|g| \leq |f|$, $f \in L$) with the Fatou property (see [BS]), is a regular class. In particular the Lebesgue space $L^p(X)$ and the Lorentz space $L^{p,q}(X)$, $0 < p, q \leq \infty$ (see (1.8)) are regular classes.

(ii) If C is a class formed by measurable sets in X containing the empty set,
$$L_C = \{f \in \mathcal{M}(X) : \{|f| > t\} \in C, \ \forall t \geq 0\}$$
is a regular class in X. To see this we observe that the conditions (i) and (ii) of Definition 1.2.1 are immediate to be checked while to prove (iii) we observe that if $0 \leq f \in L_C$, the simple functions
$$f_n = \sum_{k=0}^{n2^n-1} k2^{-n} \chi_{\{k2^{-n} < f \leq (k+1)2^{-n}\}} + n\chi_{\{f>n\}}, \quad n = 1, 2, \ldots,$$
form an increasing sequence that converges pointwise to f, whose level sets are also level sets of f and hence they belong to C.

(iii) If L is a regular class in X,
$$L^* = \{f^* : f \in L\}$$
is a regular class in $\mathbb{R}^+$.

(iv) In $\mathbb{R}^+$ the positive decreasing functions are a regular class and the same holds for the increasing functions.

(v) In $\mathbb{R}^n$ radial functions, positive and decreasing radial functions, or positive and increasing radial functions are also regular classes.

Definition 1.2.3 Let Y be a measure space, L a regular class and $T : L \to \mathcal{M}(Y)$ an operator.

(i) T is sublinear if $|T(\alpha_1 f_1 + \ldots + \alpha_n f_n)(y)| \leq |\alpha_1 T f_1(y)| + \ldots + |\alpha_n T f_n(y)|$ a.e. $y \in Y$, for every $\alpha_1, \ldots, \alpha_n \in \mathbb{R}$, and every $f_1, \ldots, f_n \in L$ such that $\alpha_1 f_1 + \ldots + \alpha_n f_n \in L$.

(ii) T is monotone if $|Tf(y)| \leq |Tg(y)|$ a.e. $y \in Y$, if $|f(x)| \leq |g(x)|$ a.e. $x \in X$, $f, g \in L$.

(iii) We say that T is order continuous if it is monotone and if for every sequence $(f_n)_n \subset L$ with $0 \leq f_n(x) \leq f_{n+1}(x) \to f(x) \in L$ a.e. $x \in X$, we have that $\lim_n |Tf_n(y)| = |Tf(y)|$ a.e. $y \in Y$.

Remark 1.2.4 It is immediate that a sublinear and monotone operator satisfies the following properties:

(i) $T(0)(y) = 0$ a.e. $y \in Y$,

(ii) $|T|f|(y)| = |Tf(y)|$ a.e. $y \in Y$, $f \in L$.

Remark 1.2.5 (i) Every maximal operator of the form

$$T^*f(x) = \sup_{T \in B} |Tf(x)|, \quad x \in X, \quad f \in L,$$

is order continuous, where, for every $x \in X$, B is a set of order continuous operators T.

(ii) In particular every integral operator $Tf(r) = \int_0^\infty k(r,t) f(t)\, dt$, $r > 0$, $f \in L \subset \mathcal{M}^+(\mathbb{R}^+)$ (with $k : \mathbb{R}^+ \times \mathbb{R}^+ \to [0, \infty))$, is order continuous. For example the Hardy operator (see (1.1)) and its conjugate

$$Qf(r) = \int_r^\infty f(t) \frac{dt}{t}.$$

(iii) Also the identity operator $f \mapsto f$ is obviously order continuous.

The following property of the sublinear and monotone operators will be fundamental.

Theorem 1.2.6 *Let $L \subset \mathcal{M}(X)$ be a regular class and $T : L \to \mathcal{M}(Y)$ a sublinear and monotone operator. Then, for every simple function $f \in L$,*

$$|Tf(y)| \leq \int_0^\infty |T\chi_{\{|f|>t\}}(y)|\, dt, \quad a.e.\ y \in Y.$$

Proof. By Remark 1.2.4 (ii) we can assume $f \geq 0$. Then,

$$f = \sum_{n=1}^N a_n \chi_{B_n},$$

with $a_1, a_2, \ldots, a_N > 0$ and $(B_j)_j$ is an increasing sequence of measurable sets in X: $B_1 \subset B_2 \subset \ldots \subset B_N$. Set $A_n = \sum_{j=n}^N a_j$, $n = 1, \ldots, N$, $A_{N+1} = 0$ and note that $\{f > t\} = \emptyset$ if $t \geq A_1$ and $\{f > t\} = B_n$ for $A_{n+1} \leq t < A_n$, $n = 1, \ldots, N$. In particular χ_{B_n} is in L (by Definition 1.2.1 (ii)) and since T is sublinear, it follows that

$$|Tf(y)| \leq \sum_{n=1}^N a_n |T\chi_{B_n}(y)|, \quad a.e.\ y \in Y. \tag{1.7}$$

On the other hand $T\chi_\emptyset(y) = 0$ a.e. $y \in Y$ (by Remark 1.2.4 (i)) and hence

$$\int_0^\infty |T\chi_{\{f>t\}}(y)|\, dt = \sum_{n=1}^N \int_{A_{n+1}}^{A_n} |T\chi_{B_n}(y)|\, dt = \sum_{n=1}^N a_n |T\chi_{B_n}(y)|,$$

since $A_n - A_{n+1} = a_n$. The last expression coincides with the right hand side of (1.7) and the theorem is proved. □

Remark 1.2.7 If the operator T is linear and positive ($Tf \geq 0$ if $f \geq 0$), the inequality in the previous theorem is, in fact, an equality when $f \geq 0$.

The following equality will be very much useful later on and it has been proved by several authors. See, for example, [HM].

Lemma 1.2.8 If $0 < p \leq 1$,

$$\sup_{f\downarrow} \frac{\left(\int_0^\infty f(t)\,dt\right)^p}{\int_0^\infty f^p(t)t^{p-1}\,dt} = p.$$

We introduce now some new spaces which play an important role in this theory, and make some important comments about its definition.

Definition 1.2.9 If $0 < p, q < \infty$, we define the Lorentz space

$$L^{p,q}(X) = \left\{ f : \|f\|_{L^{p,q}(X)} = \left(\int_0^\infty (t^{1/p} f^*(t))^q \frac{dt}{t}\right)^{1/q} < \infty \right\}. \quad (1.8)$$

If $q = \infty$ the space $L^{p,\infty}(X)$ is defined with the usual modification. When the space X is clearly understood on the context, we will simply write $\|f\|_{p,q}$.

Remark 1.2.10 (i) If $1 \leq q \leq p$, the functional $\|\cdot\|_{p,q}$ is a norm. In general, it is only a quasi-norm (see [SW]).

(ii) In the case $1 < p < q \leq \infty$ we can define

$$\|f\|_{(p,q)} = \left(\int_0^\infty (t^{1/p} f^{**}(t))^q \frac{dt}{t}\right)^{1/q}$$

(with the usual modification if $q = \infty$) which is a norm (see [BS]) and it is equivalent to the original quasi-norm:

$$\|f\|_{p,q} \leq \|f\|_{(p,q)} \leq \frac{p}{p-1} \|f\|_{p,q}, \quad f \in \mathcal{M}(X). \quad (1.9)$$

(iii) It is easy to show that $\|f\|_{p,q} = \left(\int_0^\infty (t\lambda_f^{1/p}(t))^q \frac{dt}{t}\right)^{1/q}$ (see Proposition 2.2.5).

Let us state the main result of this section.

Theorem 1.2.11 *Let $L \subset \mathcal{M}(X)$ be a regular class, $T : L \to \mathcal{M}(Y)$ an order continuous sublinear operator and $0 < q_0 \leq 1$, $0 < p_0 < \infty$. Then*

(a) If $q_0 \leq q_1 \leq p_1 < \infty$,

$$\sup_{f \in L} \frac{\|Tf\|_{L^{p_1,q_1}(Y)}}{\|f\|_{L^{p_0,q_0}(X)}} = \sup_{\chi_B \in L} \frac{\|T\chi_B\|_{L^{p_1,q_1}(Y)}}{\|\chi_B\|_{L^{p_0,q_0}(X)}}.$$

(b) If $q_0 < p_1 < q_1 \leq \infty$,

$$\sup_{f \in L} \frac{\|Tf\|_{L^{p_1,q_1}(Y)}}{\|f\|_{L^{p_0,q_0}(X)}} \leq \left(\frac{p_1}{p_1 - q_0}\right)^{1/q_0} \sup_{\chi_B \in L} \frac{\|T\chi_B\|_{L^{p_1,q_1}(Y)}}{\|\chi_B\|_{L^{p_0,q_0}(X)}}.$$

Proof. Let

$$C = \sup_{\chi_B \in L} \frac{\|T\chi_B\|_{L^{p_1,q_1}(Y)}}{\|\chi_B\|_{L^{p_0,q_0}(X)}}.$$

We have to prove that $\|Tf\|_{L^{p_1,q_1}(Y)} \leq KC\|f\|_{L^{p_0,q_0}(X)}$ for every $f \in L$ with $K = 1$ (in the case (a)) or $K = (p_1/(p_1 - q_0))^{1/q_0}$ (case (b)). By Remark 1.2.4 (ii) we can assume $f \geq 0$ and since there exists an increasing sequence of positive simple functions $(f_n)_n \subset L$ converging to f a.e., with

$$|Tf_n(y)| \leq |Tf_{n+1}(y)| \to |Tf(y)| \quad \text{a.e. } y,$$

(Definitions 1.2.1 and 1.2.3), by the monotone convergence theorem, it is enough to prove the previous inequality for $(f_n)_n$. That is, we can assume that $0 \leq f \in L$ is a simple function.

Let $p = p_1/q_0$, $q = q_1/q_0$. By Theorem 1.2.6 and Lemma 1.2.8,

$$\|Tf\|_{p_1,q_1}^{q_0} = \||Tf|^{q_0}\|_{p,q} \leq \left\| \int_0^\infty q_0 t^{q_0 - 1} |T\chi_{\{f>t\}}(\cdot)|^{q_0} \, dt \right\|_{p,q}.$$

In the case (a), $\|\cdot\|_{p,q}$ is a norm and we obtain that

$$\|Tf\|_{p_1,q_1}^{q_0} \leq \int_0^\infty q_0 t^{q_0 - 1} \||T\chi_{\{f>t\}}|^{q_0}\|_{p,q} \, dt.$$

In the case (b), $\|\cdot\|_{(p,q)}$ is a norm satisfying (1.9) and it follows that

$$\|Tf\|_{p_1,q_1}^{q_0} \leq \frac{p}{p-1} \int_0^\infty q_0 t^{q_0 - 1} \||T\chi_{\{f>t\}}|^{q_0}\|_{p,q} \, dt$$

$$= \frac{p_1}{p_1 - q_0} \int_0^\infty q_0 t^{q_0 - 1} \||T\chi_{\{f>t\}}|^{q_0}\|_{p,q} \, dt.$$

That is, in any case

$$\begin{aligned}
\|Tf\|_{p_1,q_1}^{q_0} &\le K^{q_0} \int_0^\infty q_0 t^{q_0-1} \|\|T\chi_{\{f>t\}}|^{q_0}\|_{p,q}\, dt \\
&= K^{q_0} \int_0^\infty q_0 t^{q_0-1} \|T\chi_{\{f>t\}}\|_{p_1,q_1}^{q_0}\, dt \\
&\le (KC)^{q_0} \int_0^\infty q_0 t^{q_0-1} \|\chi_{\{f>t\}}\|_{p_0,q_0}^{q_0}\, dt \\
&= (KC)^{q_0} \int_0^\infty p_0 t^{q_0-1} (\lambda_f(t))^{q_0/p_0}\, dt \\
&= (KC)^{q_0} \|f\|_{p_0,q_0}^{q_0}. \quad \square
\end{aligned}$$

Applying the previous result to the strong (i.e., diagonal) case $T: L^{p_0} \to L^{p_1}$ we obtain:

Corollary 1.2.12 *Let $L \subset \mathcal{M}(X)$ be a regular class and $T: L \to \mathcal{M}(Y)$ an order continuous sublinear operator. If $0 < p_0 \le 1$, $p_0 \le p_1 < \infty$ we have*

$$\sup_{f\in L} \frac{\|Tf\|_{L^{p_1}(Y)}}{\|f\|_{L^{p_0}(X)}} = \sup_{\chi_B \in L} \frac{\|T\chi_B\|_{L^{p_1}(Y)}}{\|\chi_B\|_{L^{p_0}(X)}}.$$

Remark 1.2.13 The range of exponents p_0, p_1 in Corollary 1.2.12 is optimal. To see this, observe:

(i) The result is not true if $p_0 > 1$. A counterexample is the Hardy operator A. Given $1 < p_0 \le p_1$, a necessary and sufficient condition to have the boundedness $A: L_{\text{dec}}^{p_0}(w_0) \to L^{p_1}(w_1)$ is (see [Sa]),

$$\begin{aligned}
W_1^{1/p_1}(r) &\le C W_0^{1/p_0}(r), \\
\left(\int_r^\infty \frac{w_1(t)}{t^{p_1}}\, dt\right)^{1/p_1} \left(\int_0^r \left(\frac{W_0(t)}{t}\right)^{-p_0'} w_0(t)\, dt\right)^{1/p_0'} &\le C. \quad (1.10)
\end{aligned}$$

It is easy to see that the condition

$$\|A\chi_{(0,r)}\|_{L^{p_1}(w_1)} \le C \|\chi_{(0,r)}\|_{L^{p_0}(w_0)}, \quad r > 0$$

obtained from Corollary 1.2.12 is equivalent to the inequalities

$$\begin{aligned}
W_1^{1/p_1}(r) &\le C_1 W_0^{1/p_0}(r), \quad r > 0, \\
\left(\int_r^\infty \left(\frac{r}{t}\right)^{p_1} w_1(t)\, dt\right)^{1/p_1} &\le C_1 W_0^{1/p_0}(r), \quad r > 0.
\end{aligned}$$

Observe that these two last inequalities are true for the weights $w_0(t) = t^{p_0-1}$, $w_1(t) = t^{p_1-1}\chi_{(1,2)}(t)$ while these weights do not satisfy (1.10).

(ii) The statement does not hold if $p_1 < p_0$. To see this, it is enough to consider $T = \mathrm{Id} : L^{p_0}_{\mathrm{dec}}(w_0) \to L^{p_1}(w_1)$. Then,

$$\sup_{f\downarrow} \frac{\int_0^\infty f^{p_1}(t) w_1(t)\, dt}{\left(\int_0^\infty f^{p_0}(t) w_0(t)\, dt\right)^{p_1/p_0}} = \sup_{g\downarrow} \frac{\int_0^\infty g(t) w_1(t)\, dt}{\left(\int_0^\infty g^{p_0/p_1}(t) w_0(t)\, dt\right)^{p_1/p_0}}. \tag{1.11}$$

The last supremum is equivalent (c.f. [Sa]), up to multiplicative constants depending only on p_0/p_1, to

$$\left(\int_0^\infty \left(\frac{W_1(t)}{W_0(t)}\right)^{p_1/(p_0-p_1)} w_1(t)\, dt\right)^{(p_0-p_1)/p_0}. \tag{1.12}$$

On the other hand, if Corollary 1.2.12 were true in this case, (1.11) would be equal to

$$\sup_{r>0} \frac{\|\chi_{(0,r)}\|_{L^{p_1}(w_1)}}{\|\chi_{(0,r)}\|_{L^{p_0}(w_0)}} = \sup_{r>0} \frac{W_1^{1/p_1}(r)}{W_0^{1/p_0}(r)}.$$

This last supremum is finite for the weights $w_i(t) = t^{\alpha_i}$, $i = 0, 1$, if $(1+\alpha_1)/p_1 = (1+\alpha_0)/p_0$, $\alpha_0, \alpha_1 > -1$, while (1.12) is always infinite in this case whatever α_1, α_2 are.

Considering the weak-type case $T : L^{p_0} \to L^{p_1,\infty}$, we have as a consequence of Theorem 1.2.11 the following statement.

Corollary 1.2.14 *Let $L \subset \mathcal{M}(X)$ be a regular class and let $T : L \to \mathcal{M}(Y)$ be an order continuous sublinear operator. If $0 < p_0 \leq 1$, $p_0 < p_1 < \infty$,*

$$\sup_{f\in L} \frac{\|Tf\|_{L^{p_1,\infty}(Y)}}{\|f\|_{L^{p_0}(X)}} \leq \left(\frac{p_1}{p_1 - p_0}\right)^{1/p_0} \sup_{\chi_B \in L} \frac{\|T\chi_B\|_{L^{p_1,\infty}(Y)}}{\|\chi_B\|_{L^{p_0}(X)}}.$$

Remark 1.2.15 The previous corollary is not true if $p_0 > 1$, as can be seen in the case

$$A : L^{p_0}_{\mathrm{dec}}(w_0) \longrightarrow L^{p_1,\infty}(w_1). \tag{1.13}$$

As was proved in [CS2] (see also [Ne1]), a necessary and sufficient condition to have (1.13) is

$$\left(\int_0^r \left(\frac{W_0(t)}{t}\right)^{-p_0'} w_0(t)\, dt\right)^{1/p_0'} W_1^{1/p_1}(r) \leq Cr, \quad r > 0, \tag{1.14}$$

$$W_1^{1/p_1}(r) \leq C W_0^{1/p_0}(r), \quad r > 0,$$

that does not coincide with the condition

$$\frac{W_1^{1/p_1}(t)}{t} \leq C \frac{W_0^{1/p_0}(r)}{r}, \quad 0 < t < r < \infty,$$

which can be obtained applying Corollary 1.2.14. For example, the two weights in Remark 1.2.13 (iii.1) satisfy this last condition but not (1.14).

A consequence of Theorem 1.2.11 and the Marcinkiewicz interpolation theorem is the following result on interpolation of operators of restricted weak-type, which is a generalization of the well known Stein-Weiss theorem (Theorem 3.15 in [SW] or Theorem 5.5 in [BS]).

Theorem 1.2.16 *Let $0 < p_0, p_1 < \infty$, $0 < q_0, q_1 \leq \infty$, $p_0 \neq p_1$, $q_0 \neq q_1$ and let us assume that $T : (L^{p_0,1} + L^{p_1,1})(X) \to \mathcal{M}(Y)$ is a sublinear order continuous operator satisfying*

$$\begin{aligned} \|T\chi_B\|_{q_0,\infty} &\leq C_0 \|\chi_B\|_{p_0,1}, \quad B \subset X, \\ \|T\chi_B\|_{q_1,\infty} &\leq C_1 \|\chi_B\|_{p_1,1}, \quad B \subset X. \end{aligned}$$

Then

$$T : L^{p,r}(X) \longrightarrow L^{q,r}(Y), \quad 0 < r \leq \infty,$$

if

$$\frac{1}{p} = \frac{1-\theta}{p_0} + \frac{\theta}{p_1}, \quad \frac{1}{q} = \frac{1-\theta}{q_0} + \frac{\theta}{q_1}, \quad 0 < \theta < 1.$$

Proof. Since $L^{s,t_0} \subset L^{s,t_1}$ if $t_0 < t_1$, there exist indices $r_0, r_1 \in (0,1)$ with $r_i < q_i$, $i = 0, 1$ and such that

$$\|T\chi_B\|_{q_i,\infty} \leq C_i \|\chi_B\|_{p_i,r_i}, \quad B \subset X, \quad i = 0, 1.$$

Theorem 1.2.11 tells us that the previous inequality holds for every function $f \in L^{p_i,r_i}$ and the result follows from the general Marcinkiewicz interpolation theorem (Theorem 5.3.2 in [BL]). □

Remark 1.2.17 (i) The norm of a characteristic function in $L^{p,q}$ does not depend (up to constants) on q. Therefore, the previous theorem remains true if we substitute the original spaces $L^{p_i,1}$ by L^{p_i,r_i} with $0 < r_i \leq \infty$, $i = 0, 1$.

(ii) In the classical result of Stein-Weiss mentioned above, the previous result is proved for a more restricted set of indices:

$$1 \leq p_0, p_1 < \infty, \qquad 1 \leq q_0, q_1 \leq \infty.$$

(iii) Corollary 1.2.12 can be also generalized to consider the case of two operators. The following result gives such extension (the proof is similar to the one given in Corollary 1.2.12, and it is also based upon some ideas found in [BPS]).

Theorem 1.2.18 *Let $L \subset \mathcal{M}(X)$ be a regular class, and let $T_i : L \longrightarrow \mathcal{M}(Y_i)$, $i = 0, 1$ be two order continuous operators. Assume that T_0 is positive and linear and T_1 is sublinear. Then, for each of the following cases:*

(a) $0 < p_0 \leq 1 \leq p_1 < \infty$,

(b) $T_1 = \mathrm{Id}$, $1 \leq p_1 < \infty$, $0 < p_0 \leq p_1$,

(c) $T_0 = \mathrm{Id}$, $0 < p_0 \leq 1$, $p_0 \leq p_1 < \infty$,

(d) $T_0 = T_1$, $0 < p_0 \leq p_1 < \infty$,

we have that

$$\sup_{f \in L} \frac{\|T_1 f\|_{L^{p_1}(Y_1)}}{\|T_0 f\|_{L^{p_0}(Y_0)}} = \sup_{\chi_B \in L} \frac{\|T_1 \chi_B\|_{L^{p_1}(Y_1)}}{\|T_0 \chi_B\|_{L^{p_0}(Y_0)}}.$$

1.3 The Hardy operator and the classes B_p

The Hardy operator A defined in (1.1) will play a fundamental role in the following chapters. In particular we shall be interested in the boundedness

$$A : L^{p_0}_{\mathrm{dec}}(w_0) \longrightarrow L^{p_1}(w_1) \tag{1.15}$$

and, also,

$$A : L^{p_0}_{\mathrm{dec}}(w_0) \longrightarrow L^{p_1, \infty}(w_1). \tag{1.16}$$

The diagonal case $A : L^p_{\mathrm{dec}}(w) \to L^p(w), p > 1$ was solved by Ariño and Muckenhoupt in [AM1]. The condition on the weight w is known as B_p. This motivates the following definition.

Definition 1.3.1 We write $w \in B_p$ if

$$A : L^p_{\text{dec}}(w) \longrightarrow L^p(w),$$

and $w \in B_{p,\infty}$ if

$$A : L^p_{\text{dec}}(w) \longrightarrow L^{p,\infty}(w).$$

Analogously we write $(w_0, w_1) \in B_{p_0,p_1}$ if the boundedness (1.15) holds and we say that $(w_0, w_1) \in B_{p_0,p_1,\infty}$ if (1.16) holds.

Remark 1.3.2 Since $L^p(w) \subset L^{p,\infty}(w)$ we have that

$$B_{p_0,p_1} \subset B_{p_0,p_1,\infty}, \quad 0 < p_0, p_1 < \infty.$$

In particular $B_p \subset B_{p,\infty}$, $0 < p < \infty$.

The characterization of the weights satisfying (1.16) is obtained applying directly Theorem 3.3 in [CS2]:

Theorem 1.3.3 *Let $0 < p_1 < \infty$. Then,*

(a) *If $p_0 > 1$ the following conditions are equivalent:*

  (i) $(w_0, w_1) \in B_{p_0,p_1,\infty}$,

  (ii) $\left(\int_0^r \left(\frac{W_0(t)}{t} \right)^{1-p_0'} dt \right)^{1/p_0'} W_1^{1/p_1}(r) \leq Cr, \quad r > 0,$

  (iii)
$$\left(\int_0^r \left(\frac{W_0(t)}{t} \right)^{-p_0'} w_0(t)\, dt \right)^{1/p_0'} W_1^{1/p_1}(r) \leq Cr,$$
$$W_1^{1/p_1}(r) \leq C W_0^{1/p_0}(r), \quad r > 0.$$

(b) *If $p_0 \leq 1$ the following conditions are equivalent:*

  (i) $(w_0, w_1) \in B_{p_0,p_1,\infty}$,

  (ii) $\dfrac{W_1^{1/p_1}(r)}{r} \leq C \dfrac{W_0^{1/p_0}(t)}{t}, \quad 0 < t < r.$

The strong boundedness $A: L^{p_0}_{\text{dec}}(w_0) \to L^{p_1}(w_1)$ is not so easy and, in fact, the case $0 < p_1 < p_0 < 1$ is still open. The result that follows characterizes the classes B_p. The proof of $(i) \Leftrightarrow (ii)$ can be found in [AM1] (case $p \geq 1$) and it is a consequence of Corollary 1.2.12 in the case $p < 1$. The equivalences with (iii) and (iv) are proved in [So].

Theorem 1.3.4 (*Ariño-Muckenhoupt, J. Soria*) *For $0 < p < \infty$ the following statements are equivalent:*

(i) $w \in B_p$.

(ii) $\displaystyle\int_r^\infty \left(\frac{r}{t}\right)^p w(t)\, dt \leq C \int_0^r w, \quad r > 0.$

(iii) $\displaystyle\int_0^r \frac{t^{p-1}}{W(t)}\, dt \leq C \frac{r^p}{W(r)}, \quad r > 0.$

(iv) $\displaystyle\int_0^r \frac{1}{W^{1/p}(t)}\, dt \leq C \frac{r}{W^{1/p}(r)}, \quad r > 0.$

The classes B_{p_0,p_1} with $p_0 \leq 1$, $p_0 \leq p_1$ can be characterized using Corollary 1.2.12 (see also [CS2]). The case $p_1, p_0 > 1$ was solved by Sawyer ([Sa]). Stepanov ([Stp2]) solved the case $0 < p_1 \leq 1 < p_0$ and Sinnamon and Stepanov ([SS]) solved the case $0 < p_1 < 1 = p_0$.

We shall also be interested in the boundedness of the discrete Hardy operator A_d acting on sequences $f = (f(n))_{n \geq 0}$ in the form

$$A_d f(n) = \frac{1}{n+1} \sum_{k=0}^n f(k), \qquad n = 0, 1, 2, \ldots$$

In the next chapter, it will be very much useful to know the boundedness of

$$A_d : \ell^p_{\text{dec}}(w) \longrightarrow \ell^{p,\infty}(w), \tag{1.17}$$

and also,

$$A_d : \ell^{p,\infty}_{\text{dec}}(w) \longrightarrow \ell^{p,\infty}(w), \tag{1.18}$$

where $w = (w(n))_n$ is a weight in $\mathbb{N}^*$, that is, a sequence of positive numbers, and $\ell^p(w)$ is the Lebesgue space L^p in $\mathbb{N}^*$, with measure $\sum_n w(n)\delta_{\{n\}}$, that is,

$$\ell^p(w) = \left\{ f = (f(n))_n \subset \mathbb{C} : \|f\|_{\ell^p(w)} = \left(\sum_{n=0}^\infty |f(n)|^p w(n)\right)^{1/p} < \infty \right\}.$$

$\ell^p_{\text{dec}}(w)$ is the class of positive and decreasing sequences f (we shall use the notation $f \downarrow$) in $\ell^p(w)$ while $\ell^{p,\infty}(w)$ is the weak version of $\ell^p(w)$ and it is defined by the semi-norm

$$\|f\|_{\ell^{p,\infty}(w)} = \sup_{t>0} t^{1/p} f_w^*(t),$$

where f_w^* is the decreasing rearrangement of $f = (f(n))_n$ in the measure space $(\mathbb{N}^*, \sum_n w(n)\delta_{\{n\}})$. Similarly as was done in $\mathbb{R}^+$, for each weight w (resp. $w_0, u, \ldots$) in $\mathbb{N}^*$ we denote by W (resp. $W_0, U, \ldots$) the sequence,

$$W(n) = \sum_{k=0}^{n} w(k), \qquad n = 0, 1, 2, \ldots \tag{1.19}$$

With this notation, it can be easily seen that

$$\|f\|_{\ell^{p,\infty}(w)} = \sup_{n \geq 0} W^{1/p}(n) f(n), \qquad f \downarrow. \tag{1.20}$$

It is now the moment to recall an important result due to E. Sawyer which will be used several times in what follows. The original proof can be seen in [Sa] and other proofs in [CS2] and [Stp2].

Theorem 1.3.5 *If $1 < p < \infty$ we have that*

$$\sup_{f \downarrow} \frac{\|f\|_{L^1(w_1)}}{\|f\|_{L^p(w_0)}} \approx \left(\int_0^\infty \left(\frac{W_1(t)}{W_0(t)} \right)^{p'-1} w_1(t)\, dt \right)^{1/p'}$$

$$\approx \left(\int_0^\infty \left(\frac{W_1(t)}{W_0(t)} \right)^{p'} w_0(t)\, dt \right)^{1/p'} + \frac{W_1(\infty)}{W_0^{1/p}(\infty)}.$$

Let us now formulate a discrete version of the above result.

Theorem 1.3.6 *Let $w = (w(n))_n$, $v = (v(n))_n$ be weights in $\mathbb{N}^*$ and let*

$$S = \sup_{f \downarrow} \frac{\sum_{n=0}^\infty f(n) v(n)}{\left(\sum_{n=0}^\infty f(n)^p w(n) \right)^{1/p}}.$$

Then,

(i) If $0 < p \leq 1$,

$$S = \sup_{n \geq 0} \frac{V(n)}{W^{1/p}(n)},$$

with W defined by (1.19) and V analogously.

(ii) If $1 < p < \infty$,

$$S \approx \left(\int_0^\infty \left(\frac{\widetilde{V}(t)}{\widetilde{W}(t)}\right)^{p'-1} \tilde{v}(t)\, dt\right)^{1/p'}$$

$$\approx \left(\int_0^\infty \left(\frac{\widetilde{V}(t)}{\widetilde{W}(t)}\right)^{p'} \tilde{w}(t)\, dt\right)^{1/p'} + \frac{\widetilde{V}(\infty)}{\widetilde{W}^{1/p}(\infty)},$$

where $\tilde{v}$ is the weight in $\mathbb{R}^+$ defined by

$$\tilde{v} = \sum_{n=0}^\infty v(n) \chi_{[n,n+1)}$$

and $\widetilde{V}(t) = \int_0^t \tilde{v}(s)\, ds$ and analogously for $\tilde{w}$ and $\widetilde{W}$.

Moreover, the constants implicit in the symbol $\approx$ only depend on p.

Proof. (i) is obtained applying Corollary 1.2.12 with $p_1 = 1$, $p_0 = p$, $X = Y = \mathbb{N}^*$, $T = \mathrm{Id}$ to the regular class L of decreasing sequences in $\mathbb{N}^*$.

(ii) can be deduced from Theorem 1.3.5 observing that

$$S = \sup_{\tilde{f}\downarrow} \frac{\int_0^\infty \tilde{f}(t)\tilde{v}(t)\, dt}{\left(\int_0^\infty \tilde{f}^p(t)\tilde{w}(t)\, dt\right)^{1/p}}. \tag{1.21}$$

To see this, note that if $f = (f(n))_n$ is a decreasing sequence in $\mathbb{N}^*$ and we define $\tilde{f} = \sum_{n=0}^\infty f(n)\chi_{[n,n+1)} \in \mathcal{M}_{\mathrm{dec}}(\mathbb{R}^+)$, it is obvious that

$$\frac{\sum_{n=0}^\infty f(n)v(n)}{\left(\sum_{n=0}^\infty f(n)^p w(n)\right)^{1/p}} = \frac{\int_0^\infty \tilde{f}(t)\tilde{v}(t)\, dt}{\left(\int_0^\infty \tilde{f}^p(t)\tilde{w}(t)\, dt\right)^{1/p}}.$$

Therefore S is less than or equal to the second member in (1.21). On the other hand, if $g \geq 0$ is a decreasing function in $\mathbb{R}^+$ and we define the decreasing sequence $f(n) = \left(\int_n^{n+1} g^p(s)\, ds\right)^{1/p}$, $n = 0, 1, \ldots$, we obtain

$$\int_0^\infty g^p(t)\tilde{w}(t)\, dt = \sum_n f(n)^p w(n),$$

while by Hölder's inequality

$$\int_0^\infty g(t)\tilde{v}(t)\,dt = \sum_n v(n)\int_n^{n+1} g(t)\,dt$$
$$\leq \sum_n v(n)\left(\int_n^{n+1} g^p(t)\,dt\right)^{1/p} = \sum_n v(n)f(n).$$

Hence,

$$\frac{\int_0^\infty g(t)\tilde{v}(t)\,dt}{\left(\int_0^\infty g^p(t)\tilde{w}(t)\,dt\right)^{1/p}} \leq \frac{\sum_{n=0}^\infty f(n)v(n)}{\left(\sum_{n=0}^\infty f(n)^p w(n)\right)^{1/p}} \leq S.$$

Thus (1.21) is proved and (ii) follows by applying Theorem 1.3.5. $\square$

The following result characterizes the boundedness (1.17).

Theorem 1.3.7

(a) If $0 < p \leq 1$ we have that $A_d : \ell^p_{\mathrm{dec}}(w) \longrightarrow \ell^{p,\infty}(w)$ if and only if

$$\frac{W^{1/p}(n)}{n+1} \leq C\,\frac{W^{1/p}(m)}{m+1}, \qquad 0 \leq m \leq n. \tag{1.22}$$

(b) If $1 < p < \infty$ the following statements are equivalent:

(i) $A_d : \ell^p_{\mathrm{dec}}(w) \longrightarrow \ell^{p,\infty}(w)$,
(ii) $\tilde{w} = \sum_{n=0}^\infty w(n)\chi_{[n,n+1)} \in B_p$,
(iii) $\sum_{k=0}^n \frac{1}{W^{1/p}(k)} \leq C\,\frac{n+1}{W^{1/p}(n)}, \quad n = 0, 1, 2, \ldots$

Proof. The boundedness of $A_d : \ell^p_{\mathrm{dec}}(w) \longrightarrow \ell^{p,\infty}(w)$ is equivalent (c.f. (1.20)) to the inequality

$$W^{1/p}(n) A_d f(n) \leq C\|f\|_{\ell^p(w)}, \qquad f \downarrow, n = 0, 1, 2, \ldots$$

Equivalently

$$\frac{W^{1/p}(n)}{n+1} \sup_{f\downarrow} \frac{\sum_{k=0}^n f(k)}{\left(\sum_{k=0}^\infty f(k)^p w(k)\right)^{1/p}} \leq C. \tag{1.23}$$

Hence, (a) is obtained applying Theorem 1.3.6 (i). Also observe that the condition (1.22) is obtained applying (1.23) to sequences of the form $f = (1, 1, 1, \ldots, 1, 0, 0, \ldots)$ (characteristic functions in $\mathbb{N}^*$) and thus it is also necessary in the case $p > 1$.

To prove (b), that is the case $p > 1$, we note, as we already did in Theorem 1.3.6, that the discrete supremum in (1.23) can be substituted by a supremum on decreasing functions in $\mathbb{R}^+$ with weights $\chi_{(0,n+1)}$ and $\tilde{w}$:

$$\sup_{g\downarrow} \frac{\int_0^{n+1} g(t)\,dt}{\left(\int_0^\infty g^p(t)\tilde{w}(t)\,dt\right)^{1/p}}.$$

Therefore (1.23) is equivalent to

$$\widetilde{W}^{1/p}(n+1)Ag(n+1) \leq C\|g\|_{L^p(\tilde{w})}, \qquad g\downarrow,\ n = 0, 1, 2, \ldots$$

In fact the previous inequality is true (with constant $2C$) substituting $n+1$ by any value $t > 0$. To see this, if $0 < t < 1$,

$$\begin{aligned}
\widetilde{W}^{1/p}(t)Ag(t) &= (w(0)t)^{1/p}\,t^{-1}\int_0^t g(s)\,ds \\
&\leq (w(0)t)^{1/p}\,t^{-1}\left(\int_0^t g^p(s)\,ds\right)^{1/p} t^{1/p'} \\
&\leq \left(\int_0^t g^p(s)\tilde{w}(s)\,ds\right)^{1/p} \leq \|g\|_{L^p(\tilde{w})},
\end{aligned}$$

and if $t \in [n, n+1)$, $n \geq 1$, we obtain, using (1.22),

$$\widetilde{W}^{1/p}(t)Ag(t) \leq \widetilde{W}^{1/p}(n+1)Ag(n) \leq 2C\widetilde{W}^{1/p}(n)Ag(n) \leq \|g\|_{L^p(\tilde{w})}.$$

Summarizing, in the case $p > 1$ the boundedness of $A_d : \ell^p_{\mathrm{dec}}(w) \longrightarrow \ell^{p,\infty}(w)$ is equivalent to the inequality

$$\widetilde{W}^{1/p}(t)Ag(t) \leq C\|g\|_{L^p(\tilde{w})}, \qquad t > 0,$$

which holds for every function $g \downarrow$ in $\mathbb{R}^+$. But this means that we have the boundedness of the continuous Hardy operator

$$A : L^p_{\mathrm{dec}}(\tilde{w}) \longrightarrow L^{p,\infty}(\tilde{w}),$$

which is equivalent (Definition 1.3.1) to $\tilde{w} \in B_{p,\infty} = B_p$ (the classes B_p and $B_{p,\infty}$ coincide if $p > 1$, see [Ne1]). We have then proved the equivalence (i) and (ii). To see the equivalence with (iii) let us first observe that this condition also implies (1.22) since if $m \leq n$,

$$\frac{m+1}{W^{1/p}(m)} \leq \sum_{k=0}^{m} \frac{1}{W^{1/p}(k)} \leq \sum_{k=0}^{n} \frac{1}{W^{1/p}(k)}$$

and by (iii), the last expression is bounded above by $C(n+1)/W^{1/p}(n)$. Hence, in the proof of (ii)$\Leftrightarrow$(iii) we can assume that (1.22) holds. With this (and taking into account the definition of $\tilde{w}$) it is immediate to show the equivalence between the condition $\tilde{w} \in B_p$ (that is (ii)), that by Theorem 1.3.4 (iv) is

$$\int_0^r \frac{1}{\widetilde{W}^{1/p}(t)}\, dt \leq C \frac{r}{\widetilde{W}^{1/p}(r)}, \qquad r > 0,$$

and condition (iii):

$$\sum_{k=0}^{n} \frac{1}{W^{1/p}(k)} \leq C \frac{n+1}{W^{1/p}(n)}, \qquad n = 0, 1, 2, \ldots$$

It is enough to discretize the integral and to note that

$$\int_n^{n+1} \frac{1}{\widetilde{W}^{1/p}(t)}\, dt \approx \frac{1}{W^{1/p}(n)}, \qquad n = 0, 1, 2, \ldots \qquad \square$$

The characterization of the boundedness (1.18) can be seen very easily and it is equivalent to the condition (b)(iii) of the previous theorem.

Theorem 1.3.8 *For $0 < p < \infty$ we have that*

$$A_d : \ell^{p,\infty}_{\mathrm{dec}}(w) \longrightarrow \ell^{p,\infty}(w)$$

if and only if

$$\sum_{k=0}^{n} \frac{1}{W^{1/p}(k)} \leq C \frac{n+1}{W^{1/p}(n)}, \qquad n = 0, 1, 2, \ldots$$

Proof. The boundedness in the statement is equivalent to the inequality

$$\|A_d f\|_{\ell^{p,\infty}(w)} \leq C, \qquad f \downarrow, \|f\|_{\ell^{p,\infty}(w)} \leq 1. \tag{1.24}$$

Now, if f is decreasing $\|f\|_{\ell^{p,\infty}(w)} = \sup_n W^{1/p}(n) f(n)$ (cf. (1.20)) and $\|f\|_{\ell^{p,\infty}(w)} \leq 1$ implies $f(n) \leq W^{-1/p}(n)$, for every n. On the other hand the sequence $W^{-1/p}$ is decreasing and with norm 1. Therefore $f = W^{-1/p}$ is the sequence that attains the maximum in the left hand side of (1.24) and the characterization will be

$$\|A_d W^{-1/p}\|_{\ell^{p,\infty}(w)} = C < \infty.$$

Taking into account that $A_d W^{-1/p}$ is decreasing and (1.20) for the norm in $\ell^{p,\infty}(w)$ of such sequences, we obtain the condition of the statement. □

Corollary 1.3.9 *If $1 < p < \infty$ the following conditions are equivalent:*

(i) $\tilde{w} = \sum_{n=0}^{\infty} w(n) \chi_{[n,n+1)} \in B_p$,

(ii) $\sum_{k=0}^{n} \frac{1}{W^{1/p}(k)} \leq C \frac{n+1}{W^{1/p}(n)}$, $\quad n = 0, 1, 2, \ldots$,

(iii) $A_d : \ell^p_{\mathrm{dec}}(w) \longrightarrow \ell^{p,\infty}(w)$,

(iv) $A_d : \ell^{p,\infty}_{\mathrm{dec}}(w) \longrightarrow \ell^{p,\infty}(w)$,

(v) $A_d : \ell^p_{\mathrm{dec}}(w) \longrightarrow \ell^p(w)$.

Proof. The first four equivalences are a consequence of the two previous theorems. On the other hand, (v) implies (iii) and (i) implies

$$A : L^p_{\mathrm{dec}}(\tilde{w}) \longrightarrow L^p(\tilde{w}).$$

It is immediate to check that this boundedness of the Hardy operator in $\mathbb{R}^+$ implies the corresponding boundedness for the discrete Hardy operator A_d, that is (v). □

Chapter 2

Lorentz spaces

2.1 Introduction

If (X, μ) is a measure space, w is a weight in $\mathbb{R}^+$ (see Definition 2.1.1 below) and $0 < p < \infty$, the Lorentz space $\Lambda_X^p(w)$ is defined as the class of all measurable functions f in X for which

$$\|f\|_{\Lambda_X^p(w)} \stackrel{\text{def.}}{=} \left(\int_0^\infty (f^*(t))^p \, w(t) \, dt \right)^{1/p} < \infty, \qquad (2.1)$$

where f^* is the nonincreasing rearrangement of f with respect to μ. These spaces were first introduced by G.G. Lorentz in [Lo2] for the case $X = (0,1)$. By choosing w properly, one obtains the spaces $L^{p,q}$ defined in (1.8). As we shall see, we also have a weak-type version denoted by $\Lambda_X^{p,\infty}(w)$.

In this chapter we are going to consider the study of analytical properties of these function spaces, giving a complete description of some previously known partial results. Already in the work of Lorentz ([Lo2]) there exists a characterization of when (2.1) defines a norm. Later, A. Haaker ([Ha]) extends this result to the case of $X = \mathbb{R}^+$ and $w \notin L^1$, and considers also the existence of a dual space. E. Sawyer ([Sa]) gives several equivalent conditions on w so that $\Lambda_{\mathbb{R}^n}^p(w)$ is a Banach space, when $p > 1$, and M.J. Carro, A. García del Amo, and J. Soria ([CGS]) study the normability in the case $p \geq 1$ and X is a nonatomic space. For these same conditions, J. Soria ([So]) characterizes the normability of the weak-type Lorentz spaces. We present a throughout account of all these properties in the general setting of resonant spaces, allowing us to also consider the discrete case $X = \mathbb{N}$ and the sequence spaces ℓ^p. These discrete spaces (which are also known as $d(w,p)$) had been previously studied for decreasing weights w (and hence $\|\cdot\|_{\Lambda^p(w)}$ is always a norm if $p \geq 1$). Here we will consider general weights and also the weak-type spaces $d^\infty(w,p)$.

This chapter is divided into several sections. After the definition and first properties of section 2.2 we consider in section 2.3 the quasi-normed Lorentz spaces (for which only minimal assumptions on the monotonicity of the primitive of w are required). Density properties for simple functions and absolute continuity of the quasi-norm are also considered. Section 2.4 is the largest and we study the duality results. In particular we characterize when $\Lambda^* = \Lambda'$ and when these are the trivial space. In section 2.5 we give necessary and sufficient conditions to have that the Lorentz spaces are Banach spaces. Finally in section 2.6 we study interpolation properties and the boundedness of certain operators.

We recall now the main definitions and notations used in what follows.

Definition 2.1.1 We denote by $\mathbb{R}^+ = (0, \infty)$. The letters w, $\tilde{w}$, w_0, ..., are used for weight functions in $\mathbb{R}^+$ (nonnegative locally integrable functions in $\mathbb{R}^+$). For a given weight w we write $W(r) = \int_0^r w(t)\,dt < \infty$, $0 \leq r < \infty$.

If (X, μ) is a measure space, $f \in \mathcal{M}(X)$, we denote the distribution function and the nonincreasing rearrangement of f as $\lambda_f(t)$ and $f^*(t)$, and if the measure $d\mu(t) = w(t)\,dt$, we will write λ_f^w and f_w^* to show the dependence on the weight w.

If $(X, \mu) = (\mathbb{R}^m, w(t)dt)$ we write $L^{p,q}(w)$ instead of $L^{p,q}(X)$.

2.2 $\Lambda_X^p(w)$ spaces

In this section (X, μ) denotes, except if otherwise mentioned, a general measure space.

Definition 2.2.1 Let w be a weight in $\mathbb{R}^+$. For $0 < p < \infty$ we define the functional $\|\cdot\|_{\Lambda_X^p(w)} : \mathcal{M}(X) \to [0, \infty]$ as

$$\|f\|_{\Lambda_X^p(w)} = \left(\int_0^\infty (f^*(t))^p w(t)\,dt\right)^{1/p}, \quad f \in \mathcal{M}(X).$$

The Lorentz space $\Lambda^p(w) = \Lambda_X^p(w)$ is the class

$$\Lambda_X^p(w) = \{f \in \mathcal{M}(X) : \|f\|_{\Lambda_X^p(w)} < \infty\}.$$

Observe that $\|f\|_{\Lambda^p_X(w)} = \|f^*\|_{L^p(w)}$. This allows us to extend the previous definition. For $0 < p, q \leq \infty$ set

$$\Lambda^{p,q}_X(w) = \{f \in \mathcal{M}(X) : \|f\|_{\Lambda^{p,q}_X(w)} = \|f^*\|_{L^{p,q}(w)} < \infty\}.$$

From now on, the notation $\Lambda^p_X(w)$ or $\Lambda^{p,q}_X(w)$ without any reference to w, means that w is a weight in $\mathbb{R}^+$ not identically zero on $(0, \mu(X))$. The symbol Λ will denote any of the spaces previously defined. Sometimes we will write Λ^p as a shorthand for $\Lambda^p_X(w)$ if X and w are clearly determined. Similarly $\Lambda(w), \Lambda_X, \Lambda^p(w)$, etc. Observe that $\Lambda^{\infty,q} = \{0\}$ if $0 < q < \infty$.

These spaces were introduced by Lorentz in [Lo1] and [Lo2] for the case $X = (0, l) \subset \mathbb{R}^+$. They are invariant under rearrangement and generalize the L^p Lebesgue spaces and $L^{p,q}$ (see Examples 2.2.3(i) and 2.2.3(ii)). The spaces $\Lambda^p(w)$ we have defined are usually called the "classical" Lorentz spaces to distinguish them from $L^{p,q}$.

Remark 2.2.2 (i) $f^*(t) = 0$ if $t \geq \mu(X)$. Hence, the behavior of the weight w in $[\mu(X), \infty)$ is irrelevant. We will assume, without loss of generality, that the weight w vanishes in $[\mu(X), \infty)$ if $\mu(X) < \infty$. Observe that, in this way, we have that $w \in L^1(\mathbb{R}^+)$ if $\mu(X) < \infty$.
 (ii) If $w \notin L^1(\mathbb{R}^+)$ then $\lim_{t \to \infty} f^*(t) = 0$ if $f \in \Lambda^{p,q}(w)$ $(p < \infty)$.
 (iii) Simple functions with finite support are in $\Lambda^{p,q}(w)$. If $w \in L^1$ then $L^\infty \subset \Lambda^{p,q}(w)$ and every simple function is in $\Lambda^{p,q}(w)$.
 (iv) Observe that $\Lambda^{p,p}(w) = \Lambda^p(w)$, $0 < p \leq \infty$.

Example 2.2.3 (i) In the case $w = 1$ we have, for $0 < p, q \leq \infty$, $\Lambda^{p,q}_X(1) = L^{p,q}(X)$. Here $W(t) = t$, $t \geq 0$.
 (ii) If $0 < p, q < \infty$, $\Lambda^q_X(t^{q/p-1}) = L^{p,q}(X)$ with equality of "norms". In this case $W(t) = \frac{p}{q} t^{q/p}$, $t \geq 0$.
 (iii) If $w = \chi_{(0,1)}$, the space $\Lambda^1_X(w) = \Lambda^1_X(\chi_{(0,1)})$ contains $L^\infty(X)$ and $W(t) = t$, $0 \leq t < 1$. In this case the functional $\|\cdot\|_{\Lambda^1}$ is a norm and the space is a "Banach function space" (see Definition 2.4.3).
 (iv) If $X = \mathbb{N}^* = \{0, 1, 2, \ldots\}$ and we consider the counting measure, then measurable functions in X are sequences $f = (f(n))_{n \geq 0} \subset \mathbb{C}$ and

$$\|f\|_{\Lambda^p_X(w)} = \Big(\sum_{n=0}^\infty (f^*(n))^p \Omega_n\Big)^{1/p},$$

where for each $n = 0, 1, 2, \ldots$, $\Omega_n = \int_n^{n+1} w(s)\, ds = W(n+1) - W(n)$. Thus, $\Lambda_X^p(w)$ depends only on the sequence of positive numbers $\Omega = (\Omega_n)_{n=0}^\infty$ and it is usually denoted as $d(\Omega, p)$. In the weak-type case,

$$\|f\|_{\Lambda_X^{p,\infty}(w)} = \sup_{n \geq 0} \Big(\sum_{k=0}^n \Omega_k\Big)^{1/p} f^*(n),$$

and we will use the symbol $d^\infty(\Omega, p)$ to denote $\Lambda_X^{p,\infty}(w)$. Given $d(\Omega, p)$ or $d^\infty(\Omega, p)$ we will assume that Ω is always a sequence of positive numbers.

It is clear that both $d(\Omega, p)$ and $d^\infty(\Omega, p)$ are always contained in $\ell^\infty(\mathbb{N}^*)$ and if $\Omega \in \ell^1(\mathbb{N}^*)$ we have in fact that $d(\Omega, p) = d^\infty(\Omega, p) = \ell^\infty(\mathbb{N}^*)$ (with equivalent norms). The only interesting case is hence $\Omega \notin \ell^1(\mathbb{N}^*)$. In this case, the space is contained in $c_0(\mathbb{N}^*)$, the space of sequences that converge to 0 and, for each $f \in d(\Omega, p)$ ($f \in d^\infty(\Omega, p)$), $(f^*(n))_{n \geq 0}$ is the nonincreasing rearrangement of the sequence $(f(n))_{n \geq 0}$.

If the sequence $\Omega = (\Omega_n)_n$ of the previous example is decreasing and $p \geq 1$, $d(\Omega, p)$ is a Banach space. This case has been studied by several authors ([Ga], [CH], [Al], [Po], [NO], [AM2], etc.). The case Ω increasing is considered in [AEP]. We will assume no a priori conditions on Ω.

The following lemma (see [CS1]) will be very useful.

Lemma 2.2.4 Let $0 < p < \infty$ and $v \geq 0$ be a measurable function in $\mathbb{R}^+$. Let $V(r) = \int_0^r v(s)\, ds$, $0 \leq r < \infty$. Then, for every decreasing function f we have that
$$\int_0^\infty f^p(s) v(s)\, ds = \int_0^\infty p t^{p-1} V(\lambda_f(t))\, dt.$$

The following result gives several equivalent expressions for the functional $\|\cdot\|_{\Lambda_X^{p,q}(w)}$. In particular, we see that it only depends on W.

Proposition 2.2.5 For $0 < p, q < \infty$ and f measurable in X,

(i) $\|f\|_{\Lambda_X^{p,q}(w)} = \Big(\int_0^\infty p t^{q-1} W^{q/p}(\lambda_f(t))\, dt\Big)^{1/q}$,

(ii) $\|f\|_{\Lambda_X^p(w)} = \Big(\int_0^\infty p t^{p-1} W(\lambda_f(t))\, dt\Big)^{1/p}$,

(iii) $\|f\|_{\Lambda_X^{p,\infty}(w)} = \sup_{t>0} t W^{1/p}(\lambda_f(t)) = \sup_{t>0} f^*(t) W^{1/p}(t)$.

Proof.
(i) Since every function and its decreasing rearrangement have the same distribution function (see [BS]) we have that,

$$W(\lambda_f(t)) = W(\lambda_{f^*}(t)) = \int_0^{\lambda_{f^*}(t)} w(s)\, ds = \int_{\{f^*>t\}} w(s)\, ds = \lambda_{f^*}^w(t).$$

By Lemma 2.2.4 and since $\lambda_{f^*}^w = (f^*)_w^*$, we obtain that

$$\left(\int_0^\infty pt^{q-1} W^{q/p}(\lambda_f(t))\, dt\right)^{1/q} = \left(\int_0^\infty pt^{q-1}(\lambda_{f^*}^w(t))^{q/p}\, dt\right)^{1/q}$$
$$= \left(\int_0^\infty t^{q/p-1} \int_0^{(f^*)_w^*(t)} qs^{q-1}\, ds\, dt\right)^{1/q}$$
$$= \left(\int_0^\infty ((f^*)_w^*(t))^q t^{q/p-1}\, dt\right)^{1/q}$$
$$= \|f\|_{\Lambda_X^{p,q}(w)}.$$

(ii) It is an immediate consequence of (i).
(iii) For the first inequality we observe that

$$\|f\|_{\Lambda_X^{p,\infty}(w)} = \|f^*\|_{L^{p,\infty}(w)} = \sup_{t>0} t(\lambda_{f^*}^w(t))^{1/p} = \sup_{t>0} tW^{1/p}(\lambda_f(t)).$$

The second inequality for characteristic functions is trivial and, by monotonicity, the general case follows. □

Remark 2.2.6 (i) If we compare 2.2.5 (i) and 2.2.5 (ii) we see that, for $q < \infty$, $\|f\|_{\Lambda_X^{p,q}(w)} = \|f\|_{\Lambda_X^q(\tilde{w})}$ where $\tilde{w}(t) = W^{q/p-1}(t)w(t)$, $0 < t < \mu(X)$. Therefore, every Lorentz space as defined here reduces to $\Lambda_X^p(w)$ and its weak version $\Lambda_X^{p,\infty}(w)$.

(ii) From 2.2.5 (iii), we deduce that $\Lambda_X^{p,\infty}(w) = \Lambda_X^{q,\infty}((q/p)\tilde{w})$ for $0 < p, q < \infty$.

For the spaces $L^{p,\infty}(X)$ it is known that the quasi-norm $\|f\|_{p,\infty}$ is, for every $q < p$, equivalent to the functional

$$\sup_{E \subset X} \|f\chi_E\|_q \mu(E)^{1/p-1/q}.$$

This is the so-called Kolmogorov condition (see e.g. [GR]). An analogous version for $\Lambda^{p,\infty}(w)$ also holds.

Proposition 2.2.7 *If $0 < q < p < \infty$ and $f \in \mathcal{M}(X)$,*

$$\|f\|_{\Lambda_X^{p,\infty}(w)} \leq \sup_{E \subset X} \|f\chi_E\|_{\Lambda_X^q(w)} W(\mu(E))^{1/p-1/q} \leq \left(\frac{p}{p-q}\right)^{1/q} \|f\|_{\Lambda_X^{p,\infty}(w)},$$

where the supremum is taken over all measurable sets $E \subset X$.

Proof. Let
$$S = \sup_{E \subset X} \|f\chi_E\|_{\Lambda_X^q(w)} W(\mu(E))^{1/p-1/q}.$$

To show the first inequality, let $E = \{|f| > t\}$, $t > 0$. Then,

$$\begin{aligned}
S &\geq \|f\chi_E\|_{\Lambda_X^q(w)} W(\mu(E))^{1/p-1/q} \\
&= \left(\int_0^\infty ((f\chi_E)^*(s))^q w(s)\, ds\right)^{1/q} W(\mu(E))^{1/p-1/q} \\
&\geq \left(t^q \int_0^{\mu(E)} w(s)\, ds\right)^{1/q} W(\mu(E))^{1/p-1/q} \\
&= t\, W(\mu(E))^{1/p} = t\, W(\lambda_f(t))^{1/p},
\end{aligned}$$

and taking the supremum in $t > 0$ we get $\|f\|_{\Lambda_X^{p,\infty}(w)} \leq S$ (c.f. 2.2.5 (iii)).

To prove the second inequality, for each $f \in \mathcal{M}(X)$, $E \subset X$, let $a = \|f\|_{\Lambda_X^{p,\infty}(w)} W(\mu(E))^{-1/p}$. Then, by 2.2.5 (iii),

$$\begin{aligned}
\|f\chi_E\|_{\Lambda_X^q(w)}^q &= \int_0^\infty qt^{q-1} W(\lambda_{f\chi_E}(t))\, dt \\
&= \int_0^a qt^{q-1} W(\lambda_{f\chi_E}(t))\, dt + \int_a^\infty qt^{q-1} W(\lambda_{f\chi_E}(t))\, dt \\
&\leq W(\mu(E)) \int_0^a qt^{q-1}\, dt + \int_a^\infty qt^{q-1} \frac{\|f\|_{\Lambda_X^{p,\infty}(w)}^p}{t^p}\, dt \\
&= \frac{p}{p-q} \|f\|_{\Lambda_X^{p,\infty}(w)}^q W(\mu(E))^{1-q/p}.
\end{aligned}$$

Hence,

$$\|f\chi_E\|_{\Lambda_X^q(w)} W(\mu(E))^{1/p-1/q} \leq \left(\frac{p}{p-q}\right)^{1/q} \|f\|_{\Lambda_X^{p,\infty}(w)}. \quad \square$$

In the following proposition we state some elementary properties for these spaces (see [BS]).

Proposition 2.2.8 *For $0 < p,q \leq \infty$ and f, g, f_k, $k \geq 1$, measurable functions in X, we have that:*

(i) $|f| \leq |g| \Rightarrow \|f\|_{\Lambda_X^{p,q}(w)} \leq \|g\|_{\Lambda_X^{p,q}(w)}$,

(ii) $\|tf\|_{\Lambda_X^{p,q}(w)} = |t|\|f\|_{\Lambda_X^{p,q}(w)}$, $t \in \mathbb{C}$,

(iii) $0 \leq f_k \leq f_{k+1} \xrightarrow[k]{} f$ a.e. $\Rightarrow \|f_k\|_{\Lambda_X^{p,q}(w)} \xrightarrow[k]{} \|f\|_{\Lambda_X^{p,q}(w)}$,

(iv) $\|\liminf_k |f_k|\|_{\Lambda_X^{p,q}(w)} \leq \liminf_k \|f_k\|_{\Lambda_X^{p,q}(w)}$,

(v) $\Lambda_X^{p,q_0}(w) \subset \Lambda_X^{p,q_1}(w)$ continuously, $0 < q_0 \leq q_1 \leq \infty$,

(vi) If $W(\mu(X)) < \infty$ then, for $0 < p_0 < p_1 \leq \infty$, we have that $\Lambda_X^{p_1,q}(w) \subset \Lambda_X^{p_0,r}(w)$, $0 < r \leq \infty$ continuously,

(vii) $\chi_E \in \Lambda_X^{p,q}(w)$ if $\mu(E) < \infty$.

The following property connects the norm convergence $\|\cdot\|_{\Lambda_X^{p,q}(w)}$ with the convergence in measure and it is related to the "completeness" property of our spaces.

Proposition 2.2.9 *Assume that $W > 0$ in $(0,\infty)$, let $\Lambda = \Lambda_X^{p,q}(w)$ be a Lorentz space and let $(f_n)_n$ be a sequence of measurable functions in X.*

(i) *If $\lim_{m,n} \|f_m - f_n\|_\Lambda = 0$ then $(f_n)_n$ is a Cauchy sequence in measure and there exists $f \in \mathcal{M}(X)$ such that $\lim \|f - f_n\|_\Lambda = 0$.*

(ii) *If $f \in \mathcal{M}(X)$ and $\lim_n \|f - f_n\|_\Lambda = 0$ then $(f_n)_n$ converges to f in measure and there exists a partial $(f_{n_k})_k$ convergent to f a.e.*

Proof. The case $q < p = \infty$ is trivial and in the case $p = q = \infty$, $\Lambda^{p,q} = L^\infty$ and the result is already known. If $p < \infty$ it is immediate, by Proposition 2.2.5, that

$$W(\lambda_f(r)) \leq C_{p,q} \frac{\|f\|_{\Lambda_X^{p,q}(w)}^p}{r^p}, \quad r > 0, \quad 0 < q \leq \infty.$$

Using the hypothesis of (i) we obtain then that $W(\lambda_{f_n - f_m}(r)) \xrightarrow[m,n]{} 0$, for every $r > 0$, which (since $W > 0$) implies $\lambda_{f_n - f_m}(r) \xrightarrow[m,n]{} 0$, $r > 0$, that is, $(f_n)_n$ is a Cauchy sequence in measure. We know that this implies the convergence in measure of $(f_n)_n$ to some measurable function f and the existence of a partial $(f_{n_k})_k$ converging to f a.e. By Proposition 2.2.8 (iv) we have that $\|f - f_n\|_\Lambda \leq \liminf_k \|f_{n_k} - f_n\|_\Lambda$ and, thus, $\lim_n \|f - f_n\|_\Lambda = 0$.

The proof of (ii) is analogue. $\square$

The functional $\|\cdot\|_\Lambda$ is not, in general, a quasi-norm and, in fact, Λ could not even be a vector space. The following lemma characterizes the quasi-normability of these spaces, which, as we will see, only depends on the weight w and on the measure space X.

Lemma 2.2.10 *If $0 < p < \infty$ and $0 < q \leq \infty$, the space $\Lambda_X^{p,q}(w)$ is quasi-normed if and only if*

$$0 < W(\mu(A \cup B)) \leq C(W(\mu(A)) + W(\mu(B))), \tag{2.2}$$

for every pair of measurable sets $A, B \subset X$ with $\mu(A \cup B) > 0$.

Proof. Sufficiency: The hypothesis implies that $W(\mu(A)) > 0$ if $\mu(A) > 0$. If $\|f\|_{\Lambda^{p,q}} = 0$, by Proposition 2.2.5, we have that $W(\lambda_f(t)) = 0$, $t > 0$, and hence, $\lambda_f(t) = 0$ for every $t > 0$, that is $f = 0$ a.e. It remains to show the quasi-triangular inequality and it suffices to prove it for positive functions. Let $0 \leq f, g \in \Lambda^{p,q}(w)$ and $t > 0$. Then $\{f + g > t\} \subset \{f > t/2\} \cup \{g > t/2\}$ and by hypothesis

$$W(\lambda_{f+g}(t)) \leq C(W(\lambda_f(t/2)) + W(\lambda_g(t/2))).$$

Since C does not depend on t, it satisfies that (Proposition 2.2.5)

$$\|f + g\|_{\Lambda^{p,q}(w)} \leq C_{p,q}(\|f\|_{\Lambda^{p,q}(w)} + \|g\|_{\Lambda^{p,q}(w)}).$$

Necessity: If A, B are two measurable sets with $\mu(A \cup B) > 0$, $\chi_{A \cup B} \leq \chi_A + \chi_B$ and since $\Lambda^{p,q}(w)$ is quasi-normed, we have that (Proposition 2.2.5),

$$\begin{aligned} 0 < C_{p,q} W^{1/p}(\mu(A \cup B)) &= \|\chi_{A \cup B}\|_{\Lambda^{p,q}(w)} \\ &\leq \|\chi_A + \chi_B\|_{\Lambda^{p,q}(w)} \\ &\leq C(\|\chi_A\|_{\Lambda^{p,q}(w)} + \|\chi_B\|_{\Lambda^{p,q}(w)}) \\ &= C'_{p,q}(W^{1/p}(\mu(A)) + W^{1/p}(\mu(B))) \end{aligned}$$

which is equivalent to the condition of the statement. $\square$

The previous result motivates the following definition.

Definition 2.2.11 Let w be a weight in $\mathbb{R}^+$. We write $W \in \Delta_2(X)$ (or $W \in \Delta_2(\mu)$) if W satisfies (2.2). If $X = \mathbb{R}$ we shall simply write $W \in \Delta_2$.

Therefore, Lemma 2.2.10 tells us that, for every $0 < p < \infty$, $0 < q \leq \infty$,

$$\Lambda_X^{p,q}(w) \text{ is quasi-normed} \Leftrightarrow W \in \Delta_2(X).$$

Proposition 2.2.12 *The following conditions are equivalent:*

(i) $W \in \Delta_2$,

(ii) $W(2r) \leq CW(r)$, $r > 0$,

(iii) $W(t+s) \leq C(W(t) + W(s))$, $t, s > 0$,

and in any of these cases, $W(t) > 0$, $t > 0$.

This proposition (whose proof is trivial) shows that $W \in \Delta_2$ is something easy to check. The condition $W \in \Delta_2$ is sufficient to have the quasi-normability of $\Lambda_X^{p,q}(w)$, independent of the measure space X and this is the content of the following theorem. The second part was proved in [CS1] and the proof of (i) is immediate.

Theorem 2.2.13 *Let $0 < p < \infty$, $0 < q \leq \infty$.*

(i) *If $W \in \Delta_2$, $\Lambda^{p,q}(w)$ is quasi-normed.*

(ii) *If X is nonatomic, $\Lambda_X^{p,q}(w)$ is quasi-normed if and only if $W \in \Delta_2$.*

That is, $\Delta_2 \subset \Delta_2(X)$ for every X and if X is nonatomic, $\Delta_2 = \Delta_2(X)$ (we are assuming that w is zero outside of $(0, \mu(X))$).

Remark 2.2.14 For the case $p = q = \infty$, it is obvious that the following properties are equivalent:

(i) $\Lambda_X^\infty(w)$ is quasi-normed.

(ii) $\Lambda_X^\infty(w) = L^\infty(X)$ (with equality of norms).

(iii) $(\Lambda_X^\infty(w), \|\cdot\|_{\Lambda_X^\infty(w)})$ is a Banach space.

(iv) $W(\mu(A)) > 0$ if $\mu(A) > 0$, $A \subset X$.

Any of these conditions hold if $W \in \Delta_2(X)$.

Let us recall the following definition that we shall use quite often in what follows.

Definition 2.2.15 A resonant measure space X ([BS]) is a σ-finite space which is nonatomic or it is a union (at most countable) of atoms with equal measure.

If X is nonatomic we have already characterized (Theorem 2.2.13) when Λ_X is quasi-normed. The following theorem completes the characterization for the case X is a resonant measure space. Its proof is immediate.

Theorem 2.2.16 *Let X be an atomic measure space whose atoms have measure $b > 0$. Then $W \in \Delta_2(X)$ if and only if*

$$W(2nb) \leq C\, W(nb), \qquad n \geq 1.$$

In particular, if $0 < p < \infty$ the spaces $d(\Omega, p)$ and $d^\infty(\Omega, p)$ (Examples 2.2.3(iv)) are quasi-normed if and only if

$$\sum_{n=1}^{2N} \Omega_{n-1} \leq C \sum_{n=1}^{N} \Omega_{n-1}, \quad N = 1, 2, \dots$$

2.3 Quasi-normed Lorentz spaces

In this section (X, μ) will be an arbitrary σ-finite measure space and the Lorentz spaces will be defined on the measure space X. We shall study the topology and some elementary properties of the quasi-normed Lorentz spaces. As we know, the condition to have that $\Lambda_X(w)$ is quasi-normed is $W \in \Delta_2(X)$ (Lemma 2.2.10).

As in every quasi-normed space, we consider in Λ the topology induced by the quasi-norm $\|\cdot\|_\Lambda$: $G \subset \Lambda$ is an open set if for every function $f \in G$ there exists $r > 0$ such that $\{g \in \Lambda : \|f - g\|_\Lambda < r\} \subset G$. We know that there exists in Λ a translation invariant distance d and an exponent $p \in (0, 1]$ so that

$$d(f, g) \leq \|f - g\|_\Lambda^p \leq 2d(f, g), \quad f, g \in \Lambda. \tag{2.3}$$

This is a common fact in every quasi-normed space (see [BL]). Inequality (2.3) tells us that the topology of Λ coincides with the one induced by the metric d and hence, Λ is a metric space. From Proposition 2.2.9 it can be also deduced that it is complete. Λ is then a quasi-Banach space. These and other facts are summarized in the following theorem. Its proof is obvious using the previous considerations and Proposition 2.2.9 (see also [BL] for property (iii) and [BS] for (iv)).

Theorem 2.3.1 *Every quasi-normed Lorentz space Λ is quasi-Banach. In particular Λ is an F-space (topological vector space which is metrizable with a translations invariant measure and complete). Every function in Λ is finite a.e. and if $(f_n)_n \subset \Lambda$,*

(i) *$(f_n)_n$ is a Cauchy sequence, if and only if $\lim_{m,n} \|f_n - f_m\|_\Lambda = 0$ and then it is also a Cauchy sequence in measure.*

(ii) *$\Lambda-\lim f_n = f$ if and only if $\lim_n \|f - f_n\|_\Lambda = 0$ and then $(f_n)_n$ converges to f in measure and there exists a partial that converges to f a.e.*

(iii) *If $p = \log(2)/\log(2C)$ where C is the constant of the quasi-norm $\|\cdot\|_\Lambda$ in Λ, there exists a p-norm in Λ equivalent to $\|\cdot\|_\Lambda^p$. If $\sum_n \|f_n\|_\Lambda^p < \infty$ the series $\sum_n f_n$ is convergent in Λ.*

(iv) *If $\Lambda_X \subset \widetilde{\Lambda}_X$ and both are quasi-normed Lorentz spaces, the embedding is continuous.*

(v) *If F is another quasi-normed space, a linear operator $T : \Lambda \to F$ is continuous if and only if $\sup_{\|f\|_\Lambda \leq 1} \|Tf\|_F < \infty$.*

2.3.1 Absolutely continuous norm

We now study the equivalent property to the dominated convergence theorem of the L^p spaces: If $\lim_n f_n(x) = f(x)$ a.e. x and $|f_n| \leq g \in L^p$ then $\|f_n - f\|_p \xrightarrow[n]{} 0$. In general, in a Banach function space (on X σ-finite) a function g as before is said to have an absolutely continuous norm. A space in which every function has absolutely continuous norm satisfies the dominated convergence theorem (see [BS]).

Definition 2.3.2 *Let $(E, \|\cdot\|)$, $E \subset \mathcal{M}(X)$, be a quasi-normed space. A function $f \in E$ is said to have absolutely continuous norm if*

$$\lim_{n \to \infty} \|f \chi_{A_n}\| = 0,$$

for every decreasing sequence of measurable sets $(A_n)_n$ with $\chi_{A_n} \to 0$ a.e. If every function in E has this property, we say that E has an absolutely continuous norm.

The connection of this property with the dominated convergence theorem is clear from the following proposition. Its proof is (except on some minor modifications) identical to the one in [BS] (pp. 14–16) for Banach function spaces and we omit it.

Proposition 2.3.3 *If Λ is a quasi-normed Lorentz space and $f \in \Lambda$, the following statements are equivalent:*

(i) f has absolutely continuous norm.

(ii) $\lim_n \|f\chi_{E_n}\|_\Lambda = 0$ if $(E_n)_n$ is a sequence of measurable sets with $\chi_{E_n} \to 0$ a.e.

(iii) $\lim_n \|f_n\|_\Lambda = 0$ if $|f_n| \leq |f|$ and $\lim_n f_n = 0$ a.e.

(iv) $\lim_n \|g - g_n\|_\Lambda = 0$ if $|g_n| \leq |f|$ and $\lim_n g_n = g$ a.e.

The following result shows that, except in the special case $\mu(X) = \infty$ and $w \in L^1$, the spaces $\Lambda_X^p(w)$ have absolutely continuous norm.

Theorem 2.3.4 *Let $0 < p < \infty$ and $\Lambda_X^p(w)$ be a quasi-normed space. Then,*

(i) If $\mu(X) < \infty$, $\Lambda_X^p(w)$ has absolutely continuous norm.

(ii) If $\mu(X) = \infty$, $\Lambda_X^p(w)$ has absolutely continuous norm if and only if $w \notin L^1$.

Proof. (i) Let us assume that $\mu(X) < \infty$ and let us see that if $0 \leq f \in \Lambda_X^p(w)$ and $(g_n)_n$ is a sequence of measurable functions satisfying

$$0 \leq g_n \leq f \in \Lambda, \quad g_n \to 0 \text{ a.e.,}$$

then $\lim_n \|g_n\|_\Lambda = 0$. Since the space is of finite measure, the pointwise convergence implies the convergence in measure and we have that $\lim_n \lambda_{g_n}(t) = 0$, $0 < t < \infty$. In particular $W(\lambda_{g_n}(t)) \to 0$, $t > 0$. Since $W(\lambda_{g_n}(t)) \leq W(\lambda_f(t))$ and $f \in \Lambda^p$, from Proposition 2.2.5(ii) and the dominated convergence theorem, it follows that $\lim_n \|g_n\|_\Lambda = 0$.

(ii) Assume that $w \notin L^1$, $0 \le f \in \Lambda_X^p(w)$ and $(g_n)_n$ is as above. The hypothesis on w implies $\lambda_f(t) < \infty$, $t > 0$ and the sets $E_k = \{f \le 1/k\}$ have complement of finite measure and, if $f_k = f\chi_{E_k}$ we have that $f_k^* \le 1/k$ and also $f_k^* \le f^*$. By the dominated convergence theorem $\|f_k\|_\Lambda \to 0$. Therefore, given $\epsilon > 0$ there exists a measurable set $E \subset X$ with $\mu(X \setminus E) < \infty$ and such that $\|f\chi_E\|_\Lambda < \epsilon$. Then the functions $g_n\chi_{X\setminus E}$ are as in (i) and $\|g_n\chi_{X\setminus E}\|_\Lambda \to 0$. We also have that

$$\limsup_n \|g_n\|_\Lambda \le C \limsup_n \left(\|g_n\chi_{X\setminus E}\|_\Lambda + \|g_n\chi_E\|_\Lambda\right) \le C\|f\chi_E\|_\Lambda \le C\epsilon.$$

And, since this is true for every $\epsilon > 0$, we obtain that $\lim_n \|g_n\|_\Lambda = 0$. This proves the sufficiency in (ii). To show the necessity, let us assume that $\mu(X) = \infty$, $w \in L^1$ and let us see that $\Lambda_X^p(w)$ has not absolutely continuous norm. In this case, since $X = \bigcup_{n=1}^\infty X_n$ with $\mu(X_n) < \infty$ for every n, the sets $E_n = \bigcup_{k=n}^\infty X_k$ satisfy $\chi_{E_n} \to 0$ a.e. and also, $\chi_{E_n}^* = 1$. Thus, $\lim_n \|\chi_{E_n}\|_{\Lambda^p} = \|w\|_1^{1/p} > 0$ and the function $1 \in \Lambda_X^p(w)$ does not have absolutely continuous norm. □

Corollary 2.3.5 *If $0 < p < \infty$ and $\Lambda_X^p(w)$ is a quasi-normed space, every function in this space which vanishes in the complement of a set of finite measure, has absolutely continuous norm.*

The analogous question in the case of the weak-type space $\Lambda_X^{p,\infty}(w)$ is solved in the following theorem, for the case when X is a resonant measure space: these spaces do not have, except in the trivial case, absolutely continuous norm.

Theorem 2.3.6 *If X is a resonant measure space, $0 < p < \infty$ and $W \in \Delta_2(X)$, $\Lambda_X^{p,\infty}(w)$ has absolutely continuous norm if and only if X is a finite union of atoms.*

Proof. If X is a finite union of atoms the proof is trivial. Hence, we shall assume that this is not the case and we consider two possibilities:

(i) If X is nonatomic, since the function $W^{-1/p}$ is decreasing and continuous in $\mathbb{R}^+$, there exists a measurable function $f \ge 0$ in X such that $f^*(t) = W^{-1/p}(t)$, $0 < t < \mu(X)$. Therefore $\|f\|_{\Lambda^{p,\infty}(w)} = \sup_t W^{1/p}(t) f^*(t) = 1$ and

$f \in \Lambda^{p,\infty}(w)$. The sets $E_n = \chi_{\{f>n\}}$, $n = 1, 2, \ldots$ form a decreasing sequence with $\chi_{E_n} \to 0$ a.e. and, however, $\|f\chi_{E_n}\|_{\Lambda^{p,\infty}} = 1$ for every n, since

$$(f\chi_{E_n})^*(t) = f^*(t)\chi_{[0,\lambda_f(n))}(t) = W^{-1/p}(t)\chi_{[0,\lambda_f(n))}(t), \quad 0 < t < \infty.$$

Hence, f does not have an absolutely continuous norm.

(ii) If $X = \bigcup_{n=0}^{\infty} X_n$ with X_n an atom of measure $b > 0$ and $X_n \cap X_m = \emptyset$, $n \neq m$, the function

$$f = \sum_{n=0}^{\infty} W^{-1/p}((n+1)b)\chi_{X_n}$$

is in $\Lambda_X^{p,\infty}(w)$ and has norm 1. The sets $E_N = \bigcup_{n=N}^{\infty} X_n$, $N = 1, 2, \ldots$ form a decreasing sequence with $\chi_{E_n} \to 0$ a.e. and, for each N,

$$(f\chi_{E_N})^*(nb) = W^{-1/p}((n+N+1)b), \quad n = 0, 1, 2, \ldots$$

Using that f^* is constant on every interval $[nb, (n+1)b)$ and Theorem 2.2.16, we obtain

$$\begin{aligned}
\|f\chi_{E_N}\|_{\Lambda^{p,\infty}} &= \sup_{t>0} W^{1/p}(t)(f\chi_{E_N})^*(t) \\
&= \sup_{n \geq 1} W^{1/p}(nb)W^{-1/p}((n+N)b) \\
&\geq (W(Nb)W^{-1}(2Nb))^{1/p} \\
&\geq C^{-1/p}.
\end{aligned}$$

Then, $\lim_N \|f\chi_{E_N}\|_{\Lambda^{p,\infty}} \neq 0$ and $\Lambda_X^{p,\infty}(w)$ does not have absolutely continuous norm. $\square$

Definition 2.3.7

$$L_0^{\infty}(X) = \{f \in L^{\infty}(X) : \mu(\{f \neq 0\}) < \infty\}.$$

We can now state a positive partial result for the spaces $\Lambda^{p,\infty}$.

Proposition 2.3.8 *If $0 < p < \infty$ and $W \in \Delta_2(X)$, every function in $L_0^{\infty}(X)$ has absolutely continuous norm in $\Lambda_X^{p,\infty}(w)$.*

Proof. If $f \in L_0^\infty$, let $Y = \{f \neq 0\} \subset X$. Then, if $(A_n)_n$ is a sequence of measurable sets with $\chi_{A_n} \to 0$ a.e., the functions $(f\chi_{A_n})_n$ are zero in the complement of Y and $\lim_n f(x)\chi_{A_n}(x) = 0$ a.e. $x \in Y$. Since Y has finite measure, it follows from Egorov's theorem, that the convergence of the previous functions is quasi-uniform. Thus, since $\epsilon > 0$, there exists a set $Y_\epsilon \subset Y$ of measure less than ϵ and such that $f\chi_{A_n} \to 0$ uniformly in $X \setminus Y_\epsilon$. Therefore, there exists $n_0 \in \mathbb{N}$ such that

$$|f(x)\chi_{A_n}(x)| < \epsilon, \quad x \in X \setminus Y_\epsilon, \ n \geq n_0.$$

Let $n \geq n_0$. Then:
(i) if $t \geq \|f\|_\infty$,

$$tW^{1/p}(\lambda_{f\chi_{A_n}}(t)) = tW^{1/p}(0) = 0,$$

(ii) if $\epsilon \leq t \leq \|f\|_\infty$,

$$tW^{1/p}(\lambda_{f\chi_{A_n}}(t)) \leq tW^{1/p}(\mu(Y_\epsilon)) \leq \|f\|_\infty W^{1/p}(\epsilon),$$

(iii) if $0 \leq t < \epsilon$,

$$tW^{1/p}(\lambda_{f\chi_{A_n}}(t)) \leq \epsilon W^{1/p}(\mu(Y))$$

and by Proposition 2.2.5 (iii) we get that

$$\|f\chi_{A_n}\|_{\Lambda_X^{p,\infty}(w)} \leq \max\{\|f\|_\infty W^{1/p}(\epsilon), \epsilon W^{1/p}(\mu(Y))\}, \quad n \geq n_0,$$

that is,

$$\limsup_n \|f\chi_{A_n}\|_{\Lambda_X^{p,\infty}(w)} \leq \max\{\|f\|_\infty W^{1/p}(\epsilon), \epsilon W^{1/p}(\mu(Y))\}.$$

Since this inequality holds for every $\epsilon > 0$ and $\lim_{t \to 0} W(t) = 0$, we conclude that

$$\lim_n \|f\chi_{A_n}\|_{\Lambda_X^{p,\infty}(w)} = 0. \quad \square$$

2.3.2 Density of the simple functions and L^∞

Let us now study the density of the simple functions and also of L^∞ in the quasi-normed Lorentz spaces. This question is solved in a positive way for the spaces Λ^p (except when $\mu(X) = \infty$, $w \in L^1$). The behavior of the density in the spaces $\Lambda^{p,\infty}$ is more irregular. It is not true here that the simple functions are dense and, in fact, not even L^∞ is always dense in $\Lambda^{p,\infty}$.

Definition 2.3.9 Let us denote by $\mathcal{S} = \mathcal{S}(X)$ the class of simple functions in X. That is
$$\mathcal{S} = \{f \in \mathcal{M}(X) : \operatorname{card}(f(X)) < \infty\}.$$
$\mathcal{S}_0$ will be the simple functions with support in a set of finite measure:
$$\mathcal{S}_0 = \mathcal{S}_0(X) = \{f \in \mathcal{S} : \mu(\{f \neq 0\}) < \infty\}.$$

It is clear that $\mathcal{S}_0 \subset L_0^\infty \subset \Lambda$ for every Lorentz space Λ.

Lemma 2.3.10 *If $f \in \mathcal{M}(X)$ there exists a sequence $(s_n)_n \subset \mathcal{S}$ satisfying*

(i) $\lim_n s_n(x) = f(x)$, $x \in X$,

(ii) $(|s_n(x)|)_n$ *is an increasing sequence and* $|s_n(x)| \leq |f(x)|$, *for every* $x \in X$,

(iii) $|f - s_n| \leq |f|$, $n \in \mathbb{N}$.

Moreover, if f is bounded then $s_n \to f$ uniformly. In particular $\mathcal{S}$ is dense in L^∞.

Theorem 2.3.11 *If $\Lambda(w)$ is a quasi-normed Lorentz space,*
$$\mathcal{S}_0 \subset L_0^\infty \subset \Lambda(w)$$
and $\mathcal{S}_0$ is dense in L_0^∞ with the topology of Λ. If $w \in L^1$,
$$\mathcal{S} \subset L^\infty \subset \Lambda(w)$$
and $\mathcal{S}$ is dense in L^∞ but, in this case, L_0^∞ is not dense in L^∞, if $\mu(X) = \infty$ (always with the topology of Λ).

Proof. The embeddings are trivial. On the other hand, if $f \in L_0^\infty$ and $E = \{f \neq 0\}$, $\mu(E) < \infty$ and if $(s_n)_n \subset \mathcal{S}$ is the sequence of the previous lemma, since $|s_n| \leq |f|$, these functions have also support in E and $(s_n)_n \subset \mathcal{S}_0$. Besides $\|f - s_n\|_{L^\infty} \to 0$ and thus,

$$\|f - s_n\|_\Lambda \leq \|\|f - s_n\|_{L^\infty} \chi_E\|_\Lambda = \|f - s_n\|_{L^\infty} \|\chi_E\|_\Lambda \to 0.$$

This proves that $\mathcal{S}_0$ is dense in L_0^∞. If $w \in L^1$, $L^\infty \subset \Lambda$ and this embedding is continuous (Theorem 2.3.1). Since $\mathcal{S}$ is dense in $(L^\infty, \|\cdot\|_{L^\infty})$ (Lemma 2.3.10), we have that $\mathcal{S}$ is dense in L^∞ with the topology of Λ. In this case $1 \in L^\infty \subset \Lambda$ and if $g \in L_0^\infty$, $|g - 1| = 1$ in a set of infinite measure if $\mu(X) = \infty$. Thus, $(1 - g)^* \geq 1$ and $\|1 - g\|_\Lambda \geq C_\Lambda > 0$. Therefore, 1 cannot be a limit in Λ of functions in L_0^∞ and this set is not dense in L^∞. $\square$

Theorem 2.3.12 *Let $0 < p < \infty$ and $W \in \Delta_2(X)$. Then,*

(i) if $w \notin L^1$, L_0^∞ is dense in $\Lambda^p(w)$,

(ii) if $w \in L^1$, L^∞ is dense in $\Lambda^p(w)$. But, in this case, $L_0^\infty(X)$ is not dense in $\Lambda_X^p(w)$, if $\mu(X) = \infty$.

Proof. Let us see that L^∞ (L_0^∞ in the case $w \notin L^1$) is dense in Λ^p. If $f \in \Lambda^p$, Proposition 2.2.5 tells us that $\lim_{t \to \infty} \lambda_f(t) = 0$. If for each $n = 1, 2, \ldots$ we define $f_n = f \chi_{\{|f| \leq n\}}$, since $f - f_n = f \chi_{\{|f| > n\}}$, then $(f - f_n)^* = f^* \chi_{[0, \lambda_f(n))}$ and we have that $\lim_n (f - f_n)^*(t) = 0$, $t > 0$. On the other hand $(f - f_n)^* \leq f^*$ and by the dominated convergence theorem, we get $\|f - f_n\|_{\Lambda^p} \to 0$. This proves that $L^\infty \cap \Lambda^p$ is dense in Λ^p. If $w \in L^1$, the first of these spaces is L^∞ and we are done. If $w \notin L^1$ we only have to see that every function in $L^\infty \cap \Lambda^p$ is limit (in Λ^p) of functions in L_0^∞. But if $g \in L^\infty \cap \Lambda^p$ and $w \notin L^1$, $\lim_{t \to \infty} g^*(t) = 0$ and the functions $g_n = g \chi_{\{|g| > g^*(n)\}} \in L_0^\infty$, $n = 1, 2, \ldots$, satisfy $(g - g_n)^* \leq g^*(n) \to 0$ and since $(g - g_n)^* \leq g^*$, the dominated convergence theorem shows that $\|g - g_n\|_{\Lambda^p} \to 0$.

It only remains to see that L_0^∞ is not dense in $\Lambda^p(w)$ if $w \in L^1$ and $\mu(X) = \infty$. But, in this case, we know by Theorem 2.3.11, that L_0^∞ is not dense in L^∞ and, hence, it can neither be dense in $\Lambda^p \supset L^\infty$. $\square$

In the weak-type case we have the following result of density.

Theorem 2.3.13 *If X is a resonant measure space, $0 < p < \infty$ and $\Lambda_X^{p,\infty}(w)$ is a quasi-normed space,*

(i) L_0^∞ is dense in $\Lambda^{p,\infty} \cap L^\infty$ if and only if $\mu(X) < \infty$ and then they coincide (the density is considered with respect to the topology of $\Lambda^{p,\infty}$).

(ii) $\Lambda^{p,\infty} \cap L^\infty$ is dense in $\Lambda^{p,\infty}$ if and only if X is atomic and then they coincide.

Hence, in both cases, if these two spaces do not coincide, the smaller one is not dense in the other.

Proof. (i) If $\mu(X) < \infty$ both spaces coincide and there is nothing to prove. If $\mu(X) = \infty$ and X is not atomic, we can take $f \in L^\infty \cap \Lambda^{p,\infty}$ with

$$f^* = W^{-1/p}(1)\chi_{[0,1)} + W^{-1/p}\chi_{[1,\infty)}.$$

If $g \in L_0^\infty$ and $E = \{g = 0\}$ then $b = \mu(X \setminus E) < \infty$ and

$$(f-g)^*(t) \geq ((f-g)\chi_E)^*(t) = (f\chi_E)^*(t) \geq f^*(b+t) = W^{-1/p}(t+b), \quad t > 1,$$

and we have that,

$$\|f - g\|_{\Lambda^{p,\infty}} = \sup_t W^{1/p}(t)(f-s)^*(t) \geq \sup_{t>1} \frac{W^{1/p}(t)}{W^{1/p}(t+b)}.$$

If $w \in L^1$ the above supremum is 1 and, in the opposite case, since $W \in \Delta_2$, $W(t+b) \leq C(W(t)+W(b))$ (cf. Theorem 2.2.13 and Proposition 2.2.12) we obtain that $\|f - g\|_{\Lambda^{p,\infty}} \geq C^{-1/p}$ and f cannot be the limit of functions in L_0^∞. This shows that L_0^∞ is not dense in $L^\infty \cap \Lambda^{p,\infty}$ in the nonatomic case. If $\mu(X) = \infty$ and X is an atomic resonant measure space, $X = \bigcup_{n=0}^\infty X_n$ with every X_n an atom of measure $b > 0$ and $X_n \cap X_m = \emptyset$, $n \neq m$, and the function

$$f = \sum_{n=0}^\infty W^{-1/p}((n+1)b)\chi_{X_n}$$

is in $\Lambda_X^{p,\infty}(w) \cap L^\infty$ and has norm 1. If $g \in L_0^\infty$, then its support has finite measure (a finite union $X_{n_1} \cup \ldots \cup X_{n_k}$ of atoms) and there exists $N \in \mathbb{N}$ such that $g = 0$ in $\bigcup_{n=N}^\infty X_n$. Hence, $(f-g)^*(nb) \geq f^*((n+N)b) = W^{-1/p}((n+N+1)b)$, $n = 0, 1, 2, \ldots$, proceeding as in the proof of Theorem 2.3.6 we can conclude that $\|f - g\|_{\Lambda^{p,\infty}} \geq C > 0$. Thus, f is not the limit of functions in L_0^∞ and consequently this set is not dense in $\Lambda^{p,\infty} \cap L^\infty$ and (i) is proved.

(ii) As before, if X is atomic both spaces coincide and there is nothing to be proved. If X is nonatomic, there exists $0 \leq f \in \mathcal{M}(X)$ with $f^* = W^{-1/p}$

in $[0, \mu(X))$. Thus, $\|f\|_{\Lambda^{p,\infty}} = 1$ and $f \in \Lambda^{p,\infty} \setminus L^\infty$. If $g \in L^\infty(X)$ with $\|g\|_\infty = b$, then $|f - g| \geq |f| - b$ and, hence, $(f - g)^*(t) \geq f^*(t) - b = W^{-1/p}(t) - b$ if $0 \leq t < \lambda_f(b) = t_0$. Therefore,

$$\|f - g\|_{\Lambda^{p,\infty}} = \sup_t W^{1/p}(t)(f - g)^*(t) \geq \sup_{0 < t < t_0} (1 - bW^{1/p}(t)) = 1,$$

since $t_0 = \lambda_f(b) > 0$. We conclude that f is not the limit in $\Lambda^{p,\infty}$ of functions in $L^\infty(X)$. This proves that $L^\infty \cap \Lambda^{p,\infty}$ is not dense in $\Lambda^{p,\infty}$ in this case. $\square$

2.4 Duality

In this section (X, μ) will be a σ-finite measure space. $d(\Omega, p)$ and its weak version $d^\infty(\Omega, p)$ will be the Lorentz spaces on $\mathbb{N}^*$ introduced in Example 2.2.3(iv). Here $\Omega = (\Omega_n)_{n=0}^\infty$ is a sequence of positive numbers $\Omega_0 \neq 0$.

We shall study the dual spaces and the associate spaces of the Lorentz spaces on X with special attention to the case when X is a resonant measure space (Definition 2.2.15). We shall describe the associate space and we shall deduce a necessary and sufficient condition to have that the dual and the associate spaces coincide. In some cases, we shall also give a description of the Banach envelope of Λ, which is of interest if this space is not a normed space. In our study, we include the sequence Lorentz spaces $d(\Omega, p)$ with Ω arbitrary. As far as we know, only the Banach case has been previously studied (that is, the case Ω decreasing).

We shall introduce the associate space generalizing the definition that can be found in [BS] in the context of Banach function spaces.

Definition 2.4.1 If $\|\cdot\| : \mathcal{M}(X) \to [0, \infty]$ is a positively homogeneous functional and $E = \{f \in \mathcal{M}(X) : \|f\| < \infty\}$, we define the associate "norm"

$$\|f\|_{E'} = \sup \left\{ \int_X |f(x)g(x)| \, d\mu(x) : \|g\| \leq 1, g \in E \right\}, \quad f \in \mathcal{M}(X).$$

The associate space of E is then $E' = \{f \in \mathcal{M}(X) : \|f\|_{E'} < \infty\}$.

Remark 2.4.2 The following properties are immediate:

(i) $\|\cdot\|_{E'}$ is subadditive and positively homogeneous and, if E contains the characteristic functions of sets of finite measure, $(E', \|\cdot\|_{E'})$ is a normed space.

(ii) If $(E, \|\cdot\|)$ is a lattice ($\|f\| \leq \|g\|$, if $|f| \leq |g|$) then E' is also a lattice. If $\|\cdot\|$ has the Fatou property, the same happens to $\|\cdot\|_{E'}$.

(iii) If we denote by $E'' = (E')'$ we have that $E \subset E''$ and $\|f\|_{E''} \leq \|f\|$ for every $f \in \mathcal{M}(X)$.

Definition 2.4.3 A Banach function space (BFS) E in X is a subspace of $\mathcal{M}(X)$ defined by $E = \{f : \|f\| < \infty\}$ where $\|\cdot\| = \|\cdot\|_E$ is a norm (called "Banach function norm") that satisfies the following properties, for $f, g, f_n \in E$, $A \subset X$ measurable (see [BS]):

(i) $\|f\| \leq \|g\|$ if $|f| \leq |g|$,

(ii) $0 \leq f_n \leq f_{n+1} \to f \Rightarrow \|f_n\| \to \|f\|$,

(iii) $\chi_A \in E$ if $\mu(A) < \infty$ and $E \neq \{0\}$,

(iv) $\int_A |f(x)| \, d\mu(x) \leq C_A \|f\|$ if $\mu(A) < \infty$.

The result that follows establishes that the associate space Λ'_X of a quasi-normed Lorentz space is a Banach function space whenever X is a resonant measure space.

Theorem 2.4.4 *If Λ_X is a quasi-normed Lorentz space on a resonant measure space X, the associate space Λ'_X is a rearrangement invariant Banach function space. For every $f \in \Lambda'_X$,*

$$\|f\|_{\Lambda'_X} = \sup \left\{ \int_0^\infty f^*(t) g^*(t) \, dt \, : \, \|g\|_{\Lambda_X} \leq 1 \right\}. \tag{2.4}$$

Moreover, a function $f \in \mathcal{M}(X)$ is in Λ'_X if and only if $\int_X |f(x) g(x)| \, d\mu(x) < \infty$ for every $g \in \Lambda_X$ and $\Lambda'_X \neq \{0\}$ if and only if $\Lambda_X \subset L^1_{\mathrm{loc}}(X)$.

Proof. Let us first prove (2.4). If $f, g \in \mathcal{M}(X)$, we have that $\int_X |f(x) g(x)| \, d\mu(x) \leq \int_0^\infty f^*(t) g^*(t) \, dt$ and thus,

$$\|f\|_{\Lambda'_X} \leq \sup \left\{ \int_0^\infty f^*(t) g^*(t) \, dt \, : \, \|g\|_{\Lambda_X} \leq 1 \right\}.$$

To prove the other inequality, we observe that for $g \in \Lambda_X$, with $\|g\|_{\Lambda_X} \leq 1$ we have, since X is resonant,

$$\int_0^\infty f^*(t)g^*(t)\,dt = \sup_{h^* = g^*} \int_X |f(x)h(x)|\,d\mu(x)$$
$$\leq \sup_{\|h\|_{\Lambda_X} \leq 1} \int_X |f(x)h(x)|\,d\mu(x) = \|f\|_{\Lambda'_X}.$$

Hence, the functional $\|\cdot\|_{\Lambda'_X}$ is rearrangement invariant.

Let us now prove that $f \in \Lambda'_X$ if and only if $\int_X |f(x)g(x)|\,d\mu(x) < \infty$ for every $g \in \Lambda_X$. The necessity is a consequence of the definition of Λ'_X. The sufficiency follows from the closed graph theorem (see [Ru]) since under the hypothesis, the linear operator $T_f(g) = fg$, $T_f : \Lambda_X \to L^1(X)$ is well defined and both are F-spaces continuously embedded in $\mathcal{M}(X)$ (from which one can immediately see that the graph is closed). T_f is then continuous and this proves (Theorem 2.3.1 (v)) that $\int_X |f(x)g(x)|\,d\mu(x) \leq C\|g\|_{\Lambda_X}$ and thus $f \in \Lambda'_X$.

Let us now see that Λ'_X is a Banach function space. That $\|\cdot\|_{\Lambda'_X}$ is a norm is immediate. Properties (i) and (ii) in the definition of BFS (Definition 2.4.3) are also trivial. Property (iv) of that definition is also very easy, since if $\mu(A) < \infty$ and $f \in \Lambda'_X$,

$$\int_A |f(x)|\,d\mu(x) \leq \|\chi_A\|_{\Lambda_X} \|f\|_{\Lambda'_X}.$$

It only remains to prove (iii). That is that $\chi_A \in \Lambda'_X$ if $\mu(A) < \infty$ and $\Lambda'_X \neq \{0\}$. In the atomic case it is trivial and if X is not atomic we have that, if $0 \neq f \in \Lambda'_X$ there exists $t > 0$ such that, if $E = \{|f| > t\}$, then $\mu(E) > 0$. Hence, $t\chi_E \leq |f|$ and $\chi_E \in \Lambda'_X$. That is, if $\Lambda'_X \neq \{0\}$ there exists $\chi_E \in \Lambda'_X$ with $b = \mu(E) > 0$. Since $\|\cdot\|_{\Lambda'_X}$ is rearrangement invariant and monotone, we have that $\chi_A \in \Lambda'$ for every measurable set A with $\mu(A) \leq b$. If $\infty > \mu(A) > b$, the above is also true since $A = \bigcup_{n=1}^N A_n$ with $\mu(A_n) \leq b$ and $\|\chi_A\|_{\Lambda'_X} \leq \sum_n \|\chi_{A_n}\|_{\Lambda'_X} < \infty$.

It only remains to prove that $\Lambda'_X \neq \{0\}$ if and only if $\Lambda_X \subset L^1_{\text{loc}}(X)$. The necessity is immediate since $\Lambda'_X \neq \{0\}$ and $\mu(A) < \infty$ then we have already seen that $\chi_A \in \Lambda'_X$ and, in particular $\int_A |f(x)|\,d\mu(x) < \infty$, $f \in \Lambda_X$. On the other hand if $\Lambda_X \subset L^1_{\text{loc}}(X)$ and $0 < \mu(A) < \infty$ we have that

$\int_X |f(x)\chi_A(x)|\, d\mu(x) < \infty$ for every $f \in \Lambda_X$ and that implies $\chi_A \in \Lambda'_X$ and $\Lambda'_X \neq \{0\}$. □

Remark 2.4.5 If in the previous theorem, we omit the condition that X is a resonant measure space we can still prove that Λ'_X is a Banach space with monotone norm and with the Fatou property. The convergence in Λ'_X implies the convergence in measure on sets of finite measure (same for the Cauchy sequences) and it is also true that $f \in \Lambda'_X$ if and only if $\int_X |f(x)g(x)|\, d\mu(x) < \infty$ for every $g \in \Lambda_X$. The last statement of the theorem is not true (in general) in this case.

The result that follows describes the associate space of Λ_X in the case X nonatomic. But first, we need to define the Lorentz space Γ.

Definition 2.4.6 If $0 < p \leq \infty$ we define

$$\Gamma^p_X(w) = \left\{ f \in \mathcal{M}(X) : \|f\|_{\Gamma^p_X(w)} = \left(\int_0^\infty (f^{**}(t))^p w(t)\, dt \right)^{1/p} < \infty \right\}.$$

The weak-type version of the previous space is

$$\Gamma^{p,\infty}_X(w) = \left\{ f \in \mathcal{M}(X) : \|f\|_{\Gamma^{p,\infty}_X(w)} = \sup_{t>0} W^{1/p}(t) f^{**}(t) < \infty \right\}.$$

The last definition can be extended in the following way. If Φ is an arbitrary function in $\mathbb{R}^+$ we write

$$\Gamma^{p,\infty}_X(d\Phi) = \left\{ f \in \mathcal{M}(X) : \|f\|_{\Gamma^{p,\infty}_X(d\Phi)} = \sup_{t>0} \Phi^{1/p}(t) f^{**}(t) < \infty \right\}.$$

We can always assume that the function Φ in the previous definition is increasing since otherwise it can be substituted by $\widetilde{\Phi}(t) = \sup_{0<s<t} \Phi(s)$, $t > 0$, which is increasing and satisfies $\|f\|_{\Gamma^{p,\infty}_X(d\Phi)} = \|f\|_{\Gamma^{p,\infty}_X(d\widetilde{\Phi})}$.

Condition (ii) of the following result is a direct consequence of Sawyer's formula stated in Theorem 1.3.5 while (i) was proved in [CS2].

Theorem 2.4.7 Let X be a nonatomic measure space and let w be an arbitrary weight in $\mathbb{R}^+$.

(i) If $0 < p \leq 1$, then
$$\Lambda_X^p(w)' = \Gamma_X^{1,\infty}(d\Phi) \qquad \text{(with equal norms)},$$
where $\Phi(t) = tW^{-1/p}(t)$, $t > 0$.

(ii) If $1 < p < \infty$ and $f \in \mathcal{M}(X)$, then
$$\|f\|_{\Lambda_X^p(w)'} \approx \left(\int_0^\infty \left(\frac{1}{W(t)} \int_0^t f^*(s)\,ds\right)^{p'} w(t)\,dt\right)^{1/p'} + \frac{\int_0^\infty f^*(t)\,dt}{W^{1/p}(\infty)}$$
$$\approx \left(\int_0^\infty \left(\frac{1}{W(t)} \int_0^t f^*(s)\,ds\right)^{p'-1} f^*(t)\,dt\right)^{1/p'}.$$

(iii) If $0 < p < \infty$, then
$$\Lambda_X^{p,\infty}(w)' = \Lambda^1(W^{-1/p}) \qquad \text{(with equal norms)}.$$

Proof. Since X is nonatomic and σ-finite, every decreasing function in $[0, \mu(X))$ equals a.e. to the decreasing rearrangement of a function in $\mathcal{M}(X)$. Besides, X is resonant and hence,
$$\|f\|_{\Lambda_X^{p,q}(w)'} = \sup_{g \in \Lambda_X^{p,q}(w)} \frac{\int_X |f(x)g(x)|\,d\mu(x)}{\|g\|_{\Lambda_X^{p,q}(w)}} = \sup_{g\downarrow} \frac{\int_0^\infty g(t)f^*(t)\,dt}{\|g\|_{L^{p,q}(w)}}.$$

The first case (i) is solved applying Corollary 1.2.12 (with $p_1 = 1$, $T = \text{Id}$) to the regular class $L = \mathcal{M}_{\text{dec}}(\mathbb{R}^+)$ or Theorem 2.12 in [CS1]. The second case (ii) is an immediate consequence of Theorem 1.3.5 (E. Sawyer). Finally, (iii) corresponds to $q = \infty$ and we have that,
$$\|f\|_{\Lambda_X^{p,\infty}(w)'} = \sup\left\{\int_0^\infty f^*(t)g(t)\,dt : \|g\|_{L^{p,\infty}(w)} = 1, \ g \downarrow\right\}.$$

Now, if $g \downarrow$, $\|g\|_{L^{p,\infty}(w)} = \sup_t W^{1/p}(t)g(t)$ and $\|g\|_{L^{p,\infty}(w)} = 1$ implies $g \leq W^{-1/p}$, and therefore
$$\|f\|_{\Lambda_X^{p,\infty}(w)'} \leq \int_0^\infty f^*(t)W^{-1/p}(t)\,dt = \|f\|_{\Lambda^1(W^{-1/p})}.$$

On the other hand $W^{-1/p}$ is decreasing, $\|W^{-1/p}\|_{L^{p,\infty}(w)} = 1$ and we have the equality. □

Remark 2.4.8 (i) If $p > 1$ and $\tilde{w}(t) = t^{p'}W^{-p'}(t)w(t)$, $t > 0$, then (ii) of the previous theorem can be stated of the following way:

$$\Lambda_X^p(w)' = \Gamma_X^{p'}(\tilde{w}), \quad \text{if } w \notin L^1,$$
$$\Lambda_X^p(w)' = \Gamma_X^{p'}(\tilde{w}) \cap L^1(X), \quad \text{if } w \in L^1.$$

It is assumed that the norm in the intersection space is the maximum of the sum of both norms and in these equalities we are assuming the equivalence of the norms.

(ii) If $w \notin L^1$, $p > 1$ and $\tilde{w}$ (as above defined) is in $B_{p'}$, the Hardy operator A satisfies the boundedness $A: L^{p'}_{\text{dec}}(\tilde{w}) \to L^{p'}(\tilde{w})$ and then,

$$\|f\|_{\Lambda_X^p(w)'} \approx \|f\|_{\Lambda_X^{p'}(\tilde{w})}, \quad f \in \mathcal{M}(X).$$

It is easy to see that this condition on the weight w is equivalent to

$$\left(\int_0^r \left(\frac{W(t)}{t}\right)^{-p'} w(t)\,dt\right)^{1/p'} W^{1/p}(r) \geq Cr, \quad r > 0, \qquad (2.5)$$

which is the opposite inequality to the condition $w \in B_{p,\infty} = B_p$ (Theorem 1.3.3). Since one of the embeddings always holds, it follows that condition (2.5) is necessary and sufficient (in the case $w \notin L^1$) to have the identity

$$\Lambda_X^p(w)' = \Lambda_X^{p'}(\tilde{w}) \qquad \text{(with equivalent norms)}.$$

(iii) If $1 < p < \infty$ the space $\Lambda_X^{p'}(w^{1-p'})$ is embedded in $\Lambda_X^p(w)'$ since, by Hölder's inequality,

$$\int_X |f(x)g(x)|\,d\mu(x) \leq \int_0^\infty (f^*(t)w^{1/p}(t))(g^*(t)w^{-1/p}(t))\,dt$$
$$\leq \|f\|_{\Lambda^p(w)} \|g\|_{\Lambda^{p'}(w^{1-p'})}.$$

If $\tilde{w}$ is as before, we have that

$$\Lambda_X^{p'}(w^{1-p'}) \subset \Lambda_X^p(w)' \subset \Gamma_X^{p'}(\tilde{w}) \subset \Lambda_X^{p'}(\tilde{w}). \qquad (2.6)$$

(iv) The comments made after Definition 2.4.6 tell us that, for every $0 < p \leq 1$,

$$\Lambda_X^p(w)' = \Gamma_X^{1,\infty}(d\Phi_p) \qquad \text{(with equal norms)},$$

where $\Phi_p(t) = \sup_{0<s<t} sW^{-1/p}(s)$, $t > 0$. Note that

$$\Phi_p(t) = \frac{t}{\inf_{0<s<t} \frac{t}{s} W^{1/p}(s)} = \frac{t}{\inf_{s>0} \max\left(1, \frac{t}{s}\right) W^{1/p}(s)} = \frac{t}{W_p(t)},$$

where $W_p(t) = \inf_{s>0} \max\left(1, \frac{t}{s}\right) W^{1/p}(s)$, $t > 0$. It is easy to check that this function is quasi-concave ([BS]). In fact, it is the biggest quasi-concave function majorized by $W^{1/p}$ and consequently it is called the greatest concave minorant of $W^{1/p}$ (see [CPSS]). Hence, we can write,

$$\|f\|_{\Lambda_X^p(w)'} = \sup_{t>0} \frac{1}{W_p(t)} \int_0^t f^*(s)\, ds, \qquad f \in \mathcal{M}(X).$$

As a first consequence of Theorem 2.4.7 we obtain the characterization of the weights w for which $\Lambda'(w) = \{0\}$.

Theorem 2.4.9 *If X is nonatomic:*

(i) If $0 < p \leq 1$, $(\Lambda_X^p(w))' \neq \{0\} \Leftrightarrow \sup_{0<t<1} \frac{t^p}{W(t)} < \infty.$

(ii) If $1 < p < \infty$, $(\Lambda_X^p(w))' \neq \{0\} \Leftrightarrow \int_0^1 \left(\frac{t}{W(t)}\right)^{p'-1} dt < \infty.$

(iii) If $0 < p < \infty$, $(\Lambda_X^{p,\infty}(w))' \neq \{0\} \Leftrightarrow \int_0^1 \frac{1}{W^{1/p}(t)}\, dt < \infty.$

Proof. Since Λ' is a BFS, it is not identically zero if and only if it contains the functions χ_E with $\mu(E) < \infty$, that is if and only if $\|\chi_E\|_{\Lambda'} < \infty$, $\mu(E) < \infty$. The conditions are then obtained applying Theorem 2.4.7. □

Remark 2.4.10 Since W is continuous in $\mathbb{R}^+$, the condition of the previous theorem only depends on the local behavior of w at 0.

Definition 2.4.11 A weight w in $\mathbb{R}^+$ is called regular (see [Re]) if it satisfies

$$\frac{W(t)}{t} \leq C w(t), \qquad t > 0,$$

with $C > 0$ independent of t. A sequence of positive numbers $(\Omega_n)_{n=0}^\infty$ is said, analogously, to be regular if

$$\frac{1}{n+1} \sum_{k=0}^n \Omega_k \leq C \Omega_n, \qquad n = 0, 1, 2, \ldots$$

Every increasing sequence and every power sequence are regular. In [Re] it is proved that if w is regular and decreasing, the spaces $\Lambda_X^{p'}(w^{1-p'})$ and $\Lambda_X^p(w)'$ coincide. In the following result, we extend this to the case of an arbitrary weight w.

Theorem 2.4.12 Let $1 < p < \infty$ and X be nonatomic. Then:

(i) If $w \notin L^1$, $\Lambda_X^p(w)' = \Lambda_X^{p'}(w^{1-p'})$ if and only if there exists $C > 0$ such that, for $r > 0$,

$$\int_0^r w^{1-p'}(t)\,dt \leq C\Big(r^{p'} W^{1-p'}(r) + \int_0^r t^{p'} W^{-p'}(t) w(t)\,dt\Big). \qquad (2.7)$$

(ii) If w is regular,

$$\Lambda_X^{p'}(w^{1-p'}) = \Lambda_X^p(w)' = \Gamma_X^{p'}(\tilde{w}) = \Lambda_X^{p'}(\tilde{w}),$$

where $\tilde{w}(t) = t^{p'} W^{-p'}(t) w(t)$, $t > 0$.

(iii) If w is increasing, $\Lambda_X^p(w)' = \Lambda_X^{p'}(w^{1-p'})$ with equality of norms.

Proof. (i) We have observed in Remark 2.4.8 that we always have $\Lambda_X^{p'}(w^{1-p'}) \subset \Lambda_X^p(w)' = \Gamma_X^{p'}(\tilde{w})$. Thus, the equality of these two spaces is equivalent to the embedding

$$\Gamma_X^{p'}(\tilde{w}) \subset \Lambda_X^{p'}(w^{1-p'}),$$

which is also equivalent to the opposite inequality for the Hardy operator,

$$\|g\|_{L^{p'}(w^{1-p'})} \leq C \|Ag\|_{L^{p'}(\tilde{w})}, \qquad g \downarrow.$$

A necessary and sufficient condition for it can be found in [CPSS] or in [Ne1]. This condition is

$$\int_0^r w^{1-p'}(t)\,dt \leq C\Big(\widetilde{W}(r) + r^{p'} \int_r^\infty t^{-p'} \tilde{w}(t)\,dt\Big),$$

which is (2.7), since $\widetilde{W}(r) = \int_0^r t^{p'} W^{-p'}(t) w(t)\,dt$, and

$$r^{p'} \int_r^\infty t^{-p'} \tilde{w}(t)\,dt = r^{p'} \int_r^\infty W^{-p'}(t) w(t)\,dt = \frac{1}{p'-1} r^{p'} W^{1-p'}(r).$$

(ii) If w is regular we have that
$$w^{1-p'}(t) = w(t)w^{-p'}(t) \leq Cw(t)t^{p'}W^{-p'}(t) = C\tilde{w}(t).$$
Hence, $\Lambda^{p'}(\tilde{w}) \subset \Lambda^{p'}(w^{1-p'})$ and by Remark 2.4.8 (iii), we conclude the equality of the four spaces in the statement.

(iii) If $f \in \mathcal{M}(X)$ and w is increasing, the function $g_0 = (f^*w^{-1})^{p'-1}$ is decreasing in $\mathbb{R}^+$ and we have that

$$\begin{aligned}
\|f\|_{\Lambda^p(w)'} &= \sup_{g\downarrow} \frac{\int_0^\infty f^*(t)g(t)\,dt}{\left(\int_0^\infty g^p(t)w(t)\,dt\right)^{1/p}} \\
&\geq \frac{\int_0^\infty f^*(t)g_0(t)\,dt}{\left(\int_0^\infty g_0^p(t)w(t)\,dt\right)^{1/p}} = \|f\|_{\Lambda^{p'}(w^{1-p'})}.
\end{aligned}$$

Since the opposite inequality is always true (Remark 2.4.8 (iii)), we get the result. $\square$

The following result describes the biassociate space of Λ^p in the case $p \leq 1$. See also [CPSS].

Theorem 2.4.13 *Let X be nonatomic and let w be a weight in $\mathbb{R}^+$. Let $0 < p \leq 1$ and W_p the greatest concave minorant of $W^{1/p}$ (Remark 2.4.8 (iv)). Then there exists a decreasing weight $\tilde{w}_p$ with $\frac{1}{2}\widetilde{W}_p \leq W_p \leq \widetilde{W}_p$ and such that $\Lambda^p_X(w)'' = \Lambda^1_X(\tilde{w}_p)$. Moreover,*

$$\frac{1}{2}\|\cdot\|_{\Lambda^1_X(\tilde{w}_p)} \leq \|\cdot\|_{\Lambda^p_X(w)''} \leq \|\cdot\|_{\Lambda^1_X(\tilde{w}_p)}.$$

Proof. Let us note that $\|g\|_{\Lambda^p(w)'} = \sup_{t>0} W^{-1/p}(t)\int_0^t g^*(s)\,ds$ (see Theorem 2.4.7) and this norm is less than or equal to 1, if and only if $\int_0^t g^*(s)\,ds \leq W^{1/p}(t)$, $t > 0$. By Lemma 2.2.4 and Proposition 2.2.5, we obtain, for $f \in \mathcal{M}(X)$,

$$\begin{aligned}
\|f\|_{\Lambda^p(w)''} &= \sup\left\{\int_0^\infty f^*(t)g^*(t)\,dt : \|g\|_{\Lambda^p(w)'} \leq 1\right\} \\
&= \sup\left\{\int_0^\infty f^*(t)\tilde{w}(t)\,dt : \tilde{w}\downarrow, \int_0^t \tilde{w}(s)\,ds \leq W^{1/p}(t), \forall t > 0\right\} \\
&= \sup\left\{\int_0^\infty \widetilde{W}(\lambda_f(t))\,dt : \widetilde{W} \in \Upsilon\right\},
\end{aligned}$$

where $\Upsilon = \{\widetilde{W} : \widetilde{w} \downarrow, \widetilde{W} \leq W^{1/p}\}$. Every function in Υ is quasi-concave and it is majorized by $W^{1/p}$. Hence, $\widetilde{W} \leq W_p$ for every $\widetilde{W} \in \Upsilon$. On the other hand W_p is quasi-concave and there exists a concave function $\widetilde{W}_p(t) = \int_0^t \tilde{w}_p(s)\,ds$, $t > 0$, with $\frac{1}{2}\widetilde{W}_p \leq W_p \leq \widetilde{W}_p$ (see [BS]). In particular $\tilde{w}_p \downarrow$ and $\frac{1}{2}\widetilde{W}_p \in \Upsilon$. It follows that

$$\frac{1}{2}\|f\|_{\Lambda_X^1(\tilde{w}_p)} = \int_0^\infty \frac{1}{2}\widetilde{W}_p(\lambda_f(t))\,dt \leq \|f\|_{\Lambda^p(w)''}$$
$$\leq \int_0^\infty \widetilde{W}_p(\lambda_f(t))\,dt = \|f\|_{\Lambda_X^1(\tilde{w}_p)}. \quad \square$$

It is convenient now to recall the definition of the discrete Hardy operator A_d that has already appeared in (1.3). This operator acts on sequences $f = (f(n))_n$ by

$$A_d f(n) = \frac{1}{n+1}\sum_{k=0}^n f(k), \quad n = 0, 1, 2, \ldots$$

We can also describe the associate space when X is resonant totally atomic. If $\mu(X) < \infty$, $\Lambda_X^{p,q}(w) = L^\infty(X)$ (with equivalent norms) and in this case it is not of interest. If $\mu(X) = \infty$ there are only, up to isomorphisms, two spaces: $d(\Omega, p)$ and $d^\infty(\Omega, p)$ (Example 2.2.3 (iv)) and the most interesting case is when $\Omega \notin \ell^1$. Up to now, the space $d(\Omega, p)'$ had been identified only in some special cases. In [Al] it is solved for $p \geq 1$ with Ω decreasing and regular. In [Po] and [NO] the case $0 < p < 1$ and $\Omega \downarrow$ is studied, while in [AEP] the associate is described under the condition $\Omega \uparrow$ unbounded (this seems to be the unique reference where the sequence Ω is not decreasing). Here, we shall identify these spaces in the most general case:

Theorem 2.4.14 Let $W_n = \sum_{k=0}^n \Omega_k$, $n = 0, 1, 2, \ldots$ Then, for every $f = (f(n))_{n=0}^\infty \subset \mathbb{C}$,

(i) If $0 < p \leq 1$,

$$\|f\|_{d(\Omega,p)'} = \sup_{n \geq 0} W_n^{-1/p} \sum_{k=0}^n f^*(k).$$

(ii) If $1 < p < \infty$ and $\Omega \notin \ell^1$,

$$C_1\|f\|_{d(\Omega,p)'} \leq \left(\sum_{n=0}^\infty (A_d f^*(n))^{p'} \widetilde{\Omega}_n\right)^{1/p'} \leq C_2\|f\|_{d(\Omega,p)'},$$

with $\widetilde{\Omega}_0 = W_0^{1-p'}$, $\widetilde{\Omega}_n = (n+1)^{p'}(W_{n-1}^{1-p'} - W_n^{1-p'})$, $n \geq 1$, and the constants C_1 and C_2 depending only on p.

(iii) If $0 < p < \infty$,
$$\|f\|_{d^\infty(\Omega,p)'} = \sum_{n=0}^\infty f^*(n) W_n^{-1/p} = \|f\|_{d(W^{-1/p},1)}.$$

Proof. By definition
$$\|f\|_{d(\Omega,p)'} = \sup_g \frac{\sum_{n=0}^\infty f(n)g(n)}{\|g\|_{d(\Omega,p)}} = \sup_{g\downarrow} \frac{\sum_{n=0}^\infty f^*(n)g(n)}{\left(\sum_{n=0}^\infty g(n)^p \Omega_n\right)^{1/p}}.$$

Then (i) can be directly deduced applying the first part of Theorem 1.3.6. Applying the second expression of the same theorem in (ii), we obtain, for $p > 1$,
$$\|f\|_{d(\Omega,p)'}^{p'} \approx \int_0^\infty \left(\frac{\widetilde{V}(t)}{\widetilde{W}(t)}\right)^{p'} \widetilde{w}(t)\,dt,$$

where, $\widetilde{v} = \sum_{n=0}^\infty f^*(n)\chi_{[n,n+1)}$, $\widetilde{w} = \sum_{n=0}^\infty \Omega_n \chi_{[n,n+1)}$, $\widetilde{V}(t) = \int_0^t \widetilde{v}(s)\,ds$, $\widetilde{W}(t) = \int_0^t \widetilde{w}(s)\,ds$, $t > 0$. Being f^* decreasing we have, for $n \geq 1$ and $t \in [n, n+1)$,
$$\frac{1}{2}(f^*(0) + \ldots + f^*(n)) \leq \widetilde{V}(t) \leq (f^*(0) + \ldots + f^*(n)).$$

On the other hand
$$\int_n^{n+1} \frac{\widetilde{w}(t)}{\widetilde{W}^{p'}(t)}\,dt = \int_0^1 \Omega_n(\Omega_0 + \ldots + \Omega_{n-1} + \Omega_n t)^{-p'}$$
$$= \frac{W_{n-1}^{1-p'} - W_n^{1-p'}}{p' - 1} = C_p \frac{\widetilde{\Omega}_n}{(n+1)^{p'}}.$$

Hence,
$$\|f\|_{d(\Omega,p)'}^{p'} \approx \int_0^1 \left(\frac{\widetilde{V}(t)}{\widetilde{W}(t)}\right)^{p'} \widetilde{w}(t)\,dt + \sum_{n=1}^\infty \int_n^{n+1} \left(\frac{\widetilde{V}(t)}{\widetilde{W}(t)}\right)^{p'} \widetilde{w}(t)\,dt$$
$$\approx f^*(0)^{p'} \Omega_0^{1-p'} + \sum_{n=1}^\infty \left(\sum_{k=0}^n f^*(k)\right)^{p'} \frac{\widetilde{\Omega}_n}{(n+1)^{p'}}$$
$$= \sum_{n=0}^\infty (A_d f^*(n))^{p'} \widetilde{\Omega}_n.$$

It only remains to prove (iii). Note that

$$\|f\|_{d^\infty(\Omega,p)'} = \sup\left\{\sum_{n=0}^\infty f^*(n)g^*(n) : \|g\|_{d^\infty(\Omega,p)} \leq 1\right\}$$

$$= \sup\left\{\sum_{n=0}^\infty f^*(n)g^*(n) : g^*(k) \leq W_k^{-1/p},\ \forall k \geq 0\right\},$$

and this supremum is attained at the sequence $W^{-1/p} = (W_n^{-1/p})_n$. Hence,

$$\|f\|_{d^\infty(\Omega,p)'} = \sum_{n=0}^\infty f^*(n)W_n^{-1/p}. \qquad \square$$

Remark 2.4.15 (i) Both $d(\Omega,p)$ and $d^\infty(\Omega,p)$ are continuously embedded in ℓ^∞ and, thus, $\ell^1 \subset d(\Omega,p)'$ (and also $\ell^1 \subset d^\infty(\Omega,p)'$) for $0 < p < \infty$. In particular, the associate space of these spaces is never trivial.

(ii) The argument used in Remark 2.4.8 (iii) is also useful here and, for $1 < p < \infty$,

$$d(\Omega^{1-p'},p') \subset d(\Omega,p)',$$

and $\|f\|_{d(\Omega,p)'} \leq \|f\|_{d(\Omega^{1-p'},p')}$ for every sequence f. If $\Omega \uparrow$ we can use the argument in the proof of Theorem 2.4.12 (iii) and conclude (as in [AEP]) that the two previous spaces (and their norms) are equal.

In [Al], Allen proves, in the case $\Omega \downarrow$, $p > 1$, that $d(\Omega,p)' = d(\Omega^{1-p'},p')$ if Ω is regular. The following theorem extends this result to the general case.

Theorem 2.4.16 *Let $1 < p < \infty$ and $\Omega \notin \ell^1$. Then $d(\Omega,p)' = d(\Omega^{1-p'},p')$ if and only if*

$$\sum_{k=0}^n \Omega_k^{1-p'} \leq C \sum_{k=0}^n (A_d\Omega(k))^{1-p'}, \qquad n = 0,1,2,\ldots$$

In particular $d(\Omega,p)' = d(\Omega^{1-p'},p')$ if Ω is regular.

Proof. The equality of the two spaces is equivalent (see Remark 2.4.15) to the embedding $d(\Omega,p)' \subset d(\Omega^{1-p'},p')$ that, by Theorem 2.4.14 (ii), holds if and only if the inequality

$$\left(\sum_{n=0}^\infty g(n)^{p'}\Omega_n^{1-p'}\right)^{1/p'} \leq C\left(\sum_{n=0}^\infty (A_dg(n))^{p'}\widetilde{\Omega}_n\right)^{1/p'} \qquad (2.8)$$

holds for every positive and decreasing sequence $g = (g(n))_n$. Here $\widetilde{\Omega}$ is the sequence defined by $\widetilde{\Omega}_0 = \Omega_0^{1-p'}$,

$$\frac{\widetilde{\Omega}_n}{(n+1)^{p'}} = \Big(\sum_{k=0}^{n-1} \Omega_k\Big)^{1-p'} - \Big(\sum_{k=0}^{n} \Omega_k\Big)^{1-p'}, \qquad n = 1, 2, 3, \ldots$$

The class of positive and decreasing sequences is regular (Definition 1.2.1) in $\mathbb{N}^*$ and we can use Theorem 1.2.18 with $T_0 = A_d$ (which is order continuous, linear and positive) to characterize (2.8). The condition is obtained "applying" the inequality to the sequences $(1, 1, \ldots, 1, 0, 0, \ldots)$ (decreasing characteristic functions in $\mathbb{N}^*$) and it is equivalent to

$$\sum_{k=0}^{n} \Omega_k^{1-p'} \lesssim \Omega_0^{1-p'} + \sum_{k=1}^{n} (k+1)^{p'}(W_{k-1}^{1-p'} - W_k^{1-p'})$$
$$+ (n+1)^{p'} \sum_{k=n+1}^{\infty} (W_{k-1}^{1-p'} - W_k^{1-p'}),$$

$n = 1, 2, \ldots$, with $W_k = \sum_{j=0}^{k} \Omega_j$. The second term is equivalent, for $n \geq 1$, to the expression

$$\Omega_0^{1-p'} + \sum_{k=1}^{n} (k+1)^{p'}(W_{k-1}^{1-p'} - W_k^{1-p'}) + (n+1)^{p'} W_n^{1-p'} \approx \sum_{k=0}^{n-1} \Big(\frac{W_k}{k+1}\Big)^{1-p'}$$
$$\approx \sum_{k=0}^{n} \Big(\frac{W_k}{k+1}\Big)^{1-p'}.$$

And the condition of the theorem is obtained.

The second assert is immediate, since every regular weight trivially satisfies the given condition. $\square$

In [NO], M. Nawrocki and A. Ortyński prove, for the case Ω decreasing, that the associate of $d(\Omega, p)$, $p \leq 1$, is ℓ^∞ if the condition $\sum_{k=0}^{n} \Omega_k \geq C(n+1)^p$ holds. We now extend this result proving that this condition also works for the general case. Moreover, this condition is also necessary.

Theorem 2.4.17 *Let $0 < p \leq 1$. Then $d(\Omega, p)' \subset \ell^\infty$, and these two spaces are equal, if and only if there exists $C > 0$ such that*

$$\sum_{k=0}^{n} \Omega_k \geq C(n+1)^p, \qquad n = 0, 1, 2, \ldots$$

Proof. By Theorem 2.4.14 (i), we have that, with $W_n = \sum_{k=0}^n \Omega_k$,

$$\|f\|_{d(\Omega,p)'} \geq W_0^{-1/p} f^*(0) = W_0^{-1/p} \|f\|_{\ell^\infty}.$$

Therefore, $d(\Omega,p)' \subset \ell^\infty$.

If the inequality of the theorem holds,

$$W_n^{-1/p} \sum_{k=0}^n f^*(k) = \frac{n+1}{W_n^{1/p}} A_d f^*(n) \leq C^{-1/p} A_d f^*(n) \leq C^{-1/p} \|f\|_{\ell^\infty},$$

for all $n = 0, 1, 2, \ldots$. Hence, $\|f\|_{d(\Omega,p)'} = \sup_n W_n^{-1/p} \sum_{k=0}^n f^*(k) \leq C^{-1/p} \|f\|_{\ell^\infty}$ and we have that $\ell^\infty = d(\Omega,p)'$.

Conversely, if $\ell^\infty \subset d(\Omega,p)'$, then $1 \in d(\Omega,p)'$ and $\sup_n \frac{n+1}{W_n^{1/p}} = \|1\|_{d(\Omega,p)'} = C < \infty$, which implies the result. $\square$

The following result describes the biassociate of $d(\Omega, p)$ in the case $p \leq 1$ in an analogous way as we did in the nonatomic case (Theorem 2.4.13). We omit here the proof.

Theorem 2.4.18 *Let $0 < p \leq 1$ and let us denote by $W_n = \sum_{k=0}^n \Omega_k$, $n = 0, 1, 2, \ldots$ Then there exists a decreasing sequence $\widetilde{\Omega} = (\widetilde{\Omega}_n)_n$ satisfying*

$$\frac{1}{2} \sum_{k=0}^n \widetilde{\Omega}_k \leq \inf_{m \geq 0} \max\left(1, \frac{n+1}{m+1}\right) W_m^{1/p} \leq \sum_{k=0}^n \widetilde{\Omega}_k,$$

and such that $d(\Omega, p)'' = d(\widetilde{\Omega}, 1)$ with equivalent norms. More precisely,

$$\frac{1}{2} \|\cdot\|_{d(\widetilde{\Omega},1)} \leq \|\cdot\|_{d(\Omega,p)''} \leq \|\cdot\|_{d(\widetilde{\Omega},1)}.$$

Let us study now the topologic dual and its connection with the associate space.

Definition 2.4.19 *If $(E, \|\cdot\|)$ is a quasi-normed space, we define the dual E^* in the usual way:*

$$E^* = \{u : E \to \mathbb{C} : u \text{ linear and continuous}\}.$$

If $u \in E^*$ we denote by $\|u\| = \|u\|_{E^*} = \sup\{|u(f)| : \|f\| \leq 1, f \in E\} \in [0, \infty)$.

The dual E^* of a quasi-normed space $(E, \|\cdot\|_E)$ is a Banach space. If E^* separates points, every $f \in E$ can be identified with the linear and continuous form $\tilde{f} : E^* \to \mathbb{C}$ such that $\tilde{f}(u) = u(f)$, $u \in E^*$. We have then a continuous injection,

$$(E, \|\cdot\|_E) \hookrightarrow (E^{**}, \|\cdot\|_{E^{**}}),$$

and the constant of this embedding is less than or equal to 1 since, for $f \in E$,

$$\|f\|_{E^{**}} = \|\tilde{f}\|_{(E^*, \|\cdot\|_{E^*})^*} = \sup_{u \in E^*} \frac{|u(f)|}{\|u\|_{E^*}} \leq \|f\|_E.$$

The Mackey topology in E associated to the dual pair (E, E^*), is defined as having as a local basis the convex envelope of the balls in E (see [Kö]). It is the finest locally convex topology in E having E^* as its topological dual. The completion $\tilde{E}$ with this topology is called Mackey completion or also Banach envelope of E. In a quasi-normed space whose dual separates points, this topology corresponds to the one induced in E by $(E^{**}, \|\cdot\|_{E^{**}})$ and the Mackey completion is then the closure of E in $(E^{**}, \|\cdot\|_{E^{**}})$.

In our case $E = \Lambda$ is a quasi-normed Lorentz space. If $f \in \Lambda'$ the application $u_f : \Lambda \to \mathbb{C}$ such that $u_f(g) = \int_X f(x)g(x)\,d\mu(x)$ is obviously linear and continuous with norm equal to $\|f\|_{\Lambda'}$. Thus, Λ' is isometrically isomorphic to a subspace of Λ^* and, in fact, we shall identify the functions of Λ' as linear and continuous forms in Λ:

$$\Lambda' \subset \Lambda^*.$$

On the other hand, Λ is continuously embedded in $(\Lambda'', \|\cdot\|_{\Lambda''})$ (Remark 2.4.2), and therefore (if Λ^* separates points) we are working with three independent topologies in Λ and two embeddings with constant 1:

$$(\Lambda, \|\cdot\|_\Lambda) \hookrightarrow (\Lambda^{**}, \|\cdot\|_{\Lambda^{**}}),$$
$$(\Lambda, \|\cdot\|_\Lambda) \hookrightarrow (\Lambda'', \|\cdot\|_{\Lambda''}).$$

Proposition 2.4.20 *If Λ is a quasi-normed Lorentz space whose dual separates points,*

$$(\Lambda, \|\cdot\|_\Lambda) \hookrightarrow (\Lambda, \|\cdot\|_{\Lambda^{**}}) \hookrightarrow (\Lambda, \|\cdot\|_{\Lambda''}),$$

and $\|f\|_{\Lambda''} \leq \|f\|_{\Lambda^{**}} \leq \|f\|_\Lambda$, $f \in \Lambda$. If in addition, $\Lambda' = \Lambda^*$, then Λ'' is isometrically identified with a subspace of Λ^{**}. In particular, $\|f\|_{\Lambda''} = \|f\|_{\Lambda^{**}}$, for every $f \in \Lambda$, in this case.

Proof. If $f \in \Lambda$, then $f \in \Lambda^{**}$ and

$$\|f\|_\Lambda \geq \|f\|_{\Lambda^{**}} = \sup_{u \in \Lambda^*} \frac{|u(f)|}{\|u\|_{\Lambda^*}} \geq \sup_{g \in \Lambda'} \frac{\int |fg|}{\|g\|_{\Lambda'}} = \|f\|_{\Lambda''}.$$

If $\Lambda' = \Lambda^*$, every continuous and linear form $u \in \Lambda^*$ is of the form $u(f) = u_g(f) = \int_X f(x)g(x)\,d\mu(x)$ with $g \in \Lambda'$. Also, to every function $f \in \Lambda''$ we can associate the linear form $\tilde{f} : \Lambda' = \Lambda^* \to \mathbb{C}$ defined by $\tilde{f}(g) = \tilde{f}(u_g) = \int_X f(x)g(x)\,d\mu(x)$ and with norm

$$\|\tilde{f}\|_{\Lambda^{**}} = \sup_{u_g \in \Lambda^*} \frac{|\tilde{f}(u_g)|}{\|u_g\|_{\Lambda^*}} = \sup_{g \in \Lambda'} \frac{\int |fg|}{\|g\|_{\Lambda'}} = \|f\|_{\Lambda''}. \quad \square$$

We shall now study, among other, the two following questions: (i) To characterize the weights w for which $\Lambda(w)^* = \{0\}$. (ii) When is $\Lambda' = \Lambda^*$? As in the case of the Lebesgue space L^p, Radon-Nikodym theorem is the key for the following result.

Proposition 2.4.21 *Let Λ be a quasi-normed Lorentz space on X. If $u \in \Lambda^*$, there exists a unique function $g \in \Lambda'$ with $\|g\|_{\Lambda'} \leq \|u\|$, satisfying*

$$u(f) = \int_X f(x)g(x)\,d\mu(x), \quad f \in L_0^\infty.$$

Proof. Let us first assume that $\mu(X) < \infty$ (in this case $\mathcal{S} \subset L^\infty \subset \Lambda$). For every $E \subset X$ let

$$\sigma(E) = u(\chi_E) \in \mathbb{C}.$$

Then σ is a complex measure in X (the fact that it is σ-additive is a consequence of the fact that $\mu(X) < \infty$), absolutely continuous with respect to the σ-finite measure μ of X, since $\mu(E) = 0$ implies $\|\chi_E\|_\Lambda = 0$ and thus $u(\chi_E) = 0$. By Radon-Nikodym theorem, there exists a function $g \in L^1(X)$ so that $\sigma(E) = \int_E g(x)\,d\mu(x)$ for every measurable set $E \subset X$. In particular (since u is linear) $u(f) = \int_X f(x)g(x)\,d\mu(x)$ for $f \in \mathcal{S} \subset \Lambda$. This also holds if $f \in L^\infty(X) \subset \Lambda$ since there exists a sequence $(s_n)_n \subset \mathcal{S} \subset \Lambda$ converging to f in Λ and also uniformly (see the proof of Theorem 2.3.11).

If $\mu(X) = \infty$, the previous argument is true in any subset $Y \subset X$ of finite measure: we define a complex measure σ_Y in Y and we obtain the existence of a function $g_Y \in L^1(Y)$ with $u(f) = \int_X f(x) g(x) \, d\mu(x)$ for every $f \in L^\infty(X)$ supported in Y. If Y_1, Y_2 are two sets of finite measure, necessarily $\int_E g_{Y_1}(x) \, d\mu(x) = \int_E g_{Y_2}(x) \, d\mu(x)$ for every measurable set $E \subset Y_1 \cap Y_2$ and it follows that $g_{Y_1}(x) = g_{Y_2}(x)$ a.e. $x \in Y_1 \cap Y_2$. Hence, since X is σ-finite we can assure the existence of a function $g \in L^1_{\text{loc}}(X)$ such that $u(f) = \int_X f(x) g(x) \, d\mu(x)$ for every $f \in L^\infty_0(X)$.

To show that $\|g\|_{\Lambda'} \leq \|u\|$, let $\alpha \in \mathcal{M}(X)$ with $|\alpha| = 1$, $\alpha g = |g|$ and let also $(X_n)_n$ be an increasing sequence of sets of finite measure with $\bigcup_n X_n = X$. If $f \in \Lambda$, we consider $f_n = |f| \chi_{\{|f| \leq n\} \cap X_n}$. Then $0 \leq f_n \leq f_{n+1} \to |f|$ and every f_n is bounded and with support in a set of finite measure. Then,

$$\int_X |f(x) g(x)| \, d\mu(x) = \lim_n \left| \int_X \alpha(x) f_n(x) g(x) \, d\mu(x) \right|$$
$$= \lim_n |u(\alpha f_n)| \leq \|u\| \lim_n \|f_n\|_\Lambda = \|u\| \|f\|_\Lambda,$$

from which $g \in \Lambda'$ and $\|g\|_{\Lambda'} \leq \|u\|$.

Finally, the uniqueness of g is also clear, since if g_1 is another function with the same properties than g, we have that $\int_E g(x) \, d\mu(x) = \int_E g_1(x) \, d\mu(x) = u(\chi_E)$ for every measurable set $E \subset X$ of finite measure, which implies that $g = g_1$. $\square$

Remark 2.4.22 If $\mu(X) = \infty$ and $w \in L^1$ every simple function, regardless of the measure of its support, is in $\Lambda_X(w)$. We can then define, for every linear form $u \in \Lambda'$,

$$\sigma(E) = u(\chi_E), \quad E \subset X \text{ measurable}.$$

But this set function, defined on the whole σ-algebra of X, is not σ-additive in general. To have this property, we need that, for every family of disjoint measurable sets $(E_n)_n$, the functions $\chi_{\bigcup_1^N E_n}$ have to converge to $\chi_{\bigcup_1^\infty E_n}$ in Λ. Contrary to what happens if $\mu(X) < \infty$, this is not true in general in this case. For example, in $\Lambda^1_{\mathbb{R}}(\chi_{(0,1)})$ the sets $E_n = (n, n+1)$ give a counterexample.

This explains why we need the restriction "support of finite measure" in the previous proposition.

Corollary 2.4.23 *If Λ is a quasi-normed Lorentz space on X, $\Lambda' = \{0\}$ if and only if every functional $u \in \Lambda^*$ is zero on $L_0^\infty(X)$. In particular, if X is resonant, Λ_X^* separates points if and only if $\Lambda_X' \neq \{0\}$.*

Corollary 2.4.24 *If Λ is a quasi-normed Lorentz space and $f \in \Lambda$ has absolutely continuous norm, then $\|f\|_{\Lambda^{**}} = \|f\|_{\Lambda''}$.*

Proof. Since f is pointwise limit of a sequence $(f_n)_n$ in L_0^∞ with $|f_n| \leq |f|$ and it has absolutely continuous norm, we have that $f = \lim_n f_n$ in Λ and, by continuity, (Proposition 2.4.20), we also have the convergence in Λ'' and in Λ^{**}. Hence, it is sufficient to prove the result assuming that $f \in L_0^\infty$. By Proposition 2.4.21, for every $u \in \Lambda^*$ there exists $g \in \Lambda'$ with $\|g\|_{\Lambda'} \leq \|u_g\|_{\Lambda^*} = \|u\|_{\Lambda^*}$ and such that $u(f) = u_g(f) = \int_X f(x)g(x)\,d\mu(x)$. Then,

$$\|f\|_{\Lambda^{**}} = \sup_{u \in \Lambda^*} \frac{|u(f)|}{\|u\|_{\Lambda^*}} = \sup_{g \in \Lambda'} \frac{|u_g(f)|}{\|u_g\|_{\Lambda^*}} = \sup_{g \in \Lambda'} \frac{\int |f(x)g(x)|\,d\mu(x)}{\|g\|_{\Lambda'}} = \|f\|_{\Lambda''}. \quad \Box$$

We can give now a representation of the dual of Λ_X^p for an arbitrary σ-finite measure space X.

Theorem 2.4.25 *If $0 < p < \infty$ and $W \in \Delta_2(X)$,*

$$\Lambda_X^p(w)^* = \Lambda_X^p(w)' \oplus \Lambda_X^p(w)^s,$$

where

$$\Lambda_X^p(w)^s = \{u \in \Lambda^p(w)^* : |u(f)| \leq C \lim_{t \to \infty} f^*(t), \forall f \in \Lambda_X^p(w)\}.$$

This last subspace is formed by functionals that are zero on the functions whose support is of finite measure and it is not zero only if $\mu(X) = \infty$ and $w \in L^1$.

Proof. That every functional in $\Lambda^p(w)^s$ is zero on functions with support of finite measure is immediate from the definition, since for such a function f we have that $\lim_{t \to \infty} f^*(t) = 0$. This also tells us that $\Lambda^p(w)' \cap \Lambda^p(w)^s = \{0\}$ since if $u = u_g$ is in the previous intersection we have that $\int_X f(x)g(x)\,d\mu(x) = u(f) = 0$ for every $f \in \Lambda^p$ with support of

finite measure. Then it follows that $\int_X f(x)g(x)\,d\mu(x) = 0$ for every $f \in \Lambda^p$ and hence $u = 0$. Let us see now the decomposition $\Lambda^* = \Lambda' + \Lambda^s$. To this end, let $u \in \Lambda^p(w)^*$. By Proposition 2.4.21, there exists $g \in \Lambda^p(w)'$ such that the continuous linear functional $u_g(f) = \int_X f(x)g(x)\,d\mu(x)$ coincides with u in L_0^∞. If $f \in \Lambda^p$ and $Y = \{f \neq 0\}$ has finite measure, then $u(f) = u_g(f)$ since u and u_g are continuous linear forms on $\Lambda_Y^p(w)$ and coincide in $L_0^\infty(Y)$, which is dense in $\Lambda_Y^p(w)$ (Theorem 2.3.12). Let $u^s = u - u_g$. Then u^s is zero on the functions of Λ^p supported in a set of finite measure. Therefore, if f any function in Λ^p and $a = \lim_{t \to \infty} f^*(t)$ and $f_n = f\chi_{\{|f| \leq a+1/n\}}$, $n = 1, 2, \ldots$, we have $u^s(f) = u^s(f_n)$ for every n (since $f - f_n$ has support of finite measure). Thus,
$$|u^s(f)| = |u^s(f_n)| \leq \|u^s\|\|f_n\|_{\Lambda^p}, \quad n = 1, 2, \ldots$$
We now have two cases: (i) $w \notin L^1$. Then $a = 0$ and $\Lambda^p(w)$ has absolutely continuous norm (Theorem 2.3.4). Since $|f_n| \leq |f|$ and $|f_n| \to 0$ a.e. we get that $\|f_n\|_{\Lambda^p} \to 0$ and $|u^s(f)| = 0 = a$. (ii) $w \in L^1$. Then $\|f_n\|_{\Lambda^p} \leq \|f_n\|_\infty \|w\|_1^{1/p} \leq (a+1/n)\|w\|_1^{1/p}$ and we have that $|u^s(f)| \leq \|u^s\|\|w\|_1^{1/p}a = Ca = C\lim_{t \to \infty} f^*(t)$. In any of these two cases, there exists $C \in (0, +\infty)$ (independent of f) such that $|u^s(f)| \leq C\lim_{t \to \infty} f^*(t)$. That is, $u^s \in \Lambda^p(w)^s$ and we have $\Lambda^* = \Lambda' \oplus \Lambda^s$.

It remains to prove that $\Lambda^p(w)^s \neq \{0\}$ if and only if $\mu(X) = \infty$ and $w \in L^1$. If $\mu(X) < \infty$ or if $w \notin L^1$ every function $f \in \Lambda^p(w)$ satisfies $\lim_{t \to \infty} f^*(t) = 0$ and therefore, for $u \in \Lambda^s$, we get $u(f) = 0$, for every $f \in \Lambda^p$. That is $u = 0$ or equivalently $\Lambda^s = \{0\}$. If $\mu(X) = \infty$ and $w \in L^1$, the functional $p(f) = \lim_{t \to \infty} f^*(t)$, $p : \Lambda^p \to [0, +\infty)$, is a seminorm:

(i) $\quad p(\lambda f) = |\lambda p(f)|$ (obvious),

(ii) $\quad p(f+g) = \lim_{t \to \infty}(f+g)^*(t) \leq \lim_t (f^*(t/2) + g^*(t/2)) = p(f) + p(g)$.

Since $p(1) = 1$ (the constant function 1 is in $\Lambda^p(w)$ in this case) p is not zero and there exists a nonzero linear form u on Λ^p satisfying

$$|u(f)| \leq p(f) = \lim_{t \to \infty} f^*(t), \quad \forall f \in \Lambda^p \tag{2.9}$$

(see [Ru] for example). In particular, u is continuous since

$$\|f\|_{\Lambda^p(w)} = \left(\int_0^\infty (f^*(t))^p w(t)\,dt\right)^{1/p} \geq p(f)\|w\|_1^{1/p}, \quad \forall f,$$

and it follows that $|u(f)| \leq p(f) \leq C\|f\|_{\Lambda^p}$, for every f. Finally, (2.9) tells us that $u \in \Lambda^s$ and hence $\Lambda^s \neq \{0\}$. □

We shall state two immediate consequences of the last theorem that solve, in the case $\Lambda = \Lambda^p$, the two questions about duality we were looking for.

Corollary 2.4.26 *Let $0 < p < \infty$ and $W \in \Delta_2(X)$.*

(i) *If $\mu(X) < \infty$ or $w \notin L^1$,*

$$\Lambda_X^p(w)' = \Lambda_X^p(w)^*.$$

In particular $d(\Omega, p)^* = d(\Omega, p)'$ if $\Omega \notin \ell^1$ and $d(\Omega, p)$ is quasi-normed.

(ii) *If $\mu(X) = \infty$ and $w \in L^1$,*

$$\Lambda_X^p(w)' \subsetneq \Lambda_X^p(w)^*.$$

In particular $\Lambda_X^p(w)^* \neq \{0\}$ in this case.

As a consequence of this last result, Theorems 2.4.14, 2.4.16, 2.4.17 and Remark 2.4.15, remain true (in the case $d(\Omega, p)$ quasi-normed, $\Omega \notin \ell^1$) if we substitute $d(\Omega, p)'$ by $d(\Omega, p)^*$.

Corollary 2.4.27 *Let X be a nonatomic measure space and let $W \in \Delta_2$.*

(i) *If $0 < p \leq 1$, $\Lambda_X^p(w)^* = \{0\}$ if and only if the two following conditions hold:*

i.1) $\quad \mu(X) < \infty$ or else $\mu(X) = \infty$ and $w \notin L^1$, and

i.2) $\quad \sup\limits_{0 < t < 1} \dfrac{t^p}{W(t)} = \infty.$

(ii) *If $1 < p < \infty$, $\Lambda_X^p(w)^* = \{0\}$ if and only if it holds:*

ii.1) $\quad \mu(X) < \infty$ or else $\mu(X) = \infty$ and $w \notin L^1$, and

ii.2) $\quad \displaystyle\int_0^1 \left(\dfrac{t}{W(t)}\right)^{p'-1} dt = \infty.$

The spaces $d(\Omega,p)$ have been studied up to now in the case $\Omega\downarrow$. As we shall see, this condition asserts (if $p\geq 1$) that $\|\cdot\|_{d(\Omega,p)}$ is a norm. In the case $\Omega\downarrow, p<1$ there has been interest in finding the Banach envelope of $d(\Omega,p)$ (see for example [Po] and [NO]). The following theorem solves this question in the general case of $w\notin L^1$.

Theorem 2.4.28 *Let $0<p\leq 1$.*

(i) *Let X be nonatomic and let us assume that*

  (a) $W\in\Delta_2$,

  (b) $w\notin L^1$ or $\mu(X)<\infty$,

  (c) $\sup_{0<t<1}\frac{t^p}{W(t)}<\infty$.

Then, there exists a decreasing weight $\tilde{w}$ with

$$\widetilde{W}(t)\approx\inf_{s>0}\max(1,t/s)W^{1/p}(s), t>0,$$

and such that the Mackey topology in $\Lambda_X^p(w)$ is the one induced by the norm $\|\cdot\|_{\Lambda_X^1(\tilde{w})}$. The Banach envelope of $\Lambda^p(w)$ is $\Lambda^1(\tilde{w})$ if $\tilde{w}\notin L^1$, and it is the space

$$\Lambda^1(\tilde{w})_0=\{f\in\Lambda^1(\tilde{w}):\lim_{t\to\infty}f^*(t)=0\}$$

otherwise.

(ii) *If $\Omega\notin\ell^1$ and the space $d(\Omega,p)$ is quasi-normed, there exists a decreasing sequence $\tilde{\Omega}=(\tilde{\Omega}_n)_n$ such that*

$$\sum_{k=0}^n\tilde{\Omega}_k\approx\sup_{m\geq 0}\max\left(1,\frac{n+1}{m+1}\right)\left(\sum_{k=0}^m\Omega_k\right)^{1/p}, n=0,1,2,\ldots,$$

and the norm $\|\cdot\|_{d(\tilde{\Omega},1)}$ induces the Mackey topology in $d(\Omega,p)$. If $\tilde{\Omega}\notin\ell^1$ the Mackey completion of $d(\Omega,p)$ is $d(\tilde{\Omega},1)$, and this space is $c_0=c_0(\mathbb{N}^*)$ otherwise.

Proof. By Corollary 2.4.26 the dual and the associate space of $\Lambda=\Lambda_X^p(w)$ ($\Lambda=d(\Omega,p)$ in the case (ii)) coincide, and moreover $\Lambda^p(w)'=\Lambda^p(w)^*\neq\{0\}$ by Theorem 2.4.9. It follows then from Proposition 2.4.20 that the Mackey topology in Λ (the topology of the bidual norm) is the one induced by the norm of the biassociate space, and since Λ'' is complete, the Mackey completion $\widetilde{\Lambda}$ is contained in Λ''. By Theorem 2.4.13 (Theorem 2.4.18 in the case (ii)), $\Lambda'' = \Lambda^1(\widetilde{w})$ ($\Lambda'' = d(\widetilde{\Omega}, 1)$ resp.). If $\widetilde{w} \notin L^1$ ($\widetilde{\Omega} \notin \ell^1$), Theorem 2.3.12 tells us that Λ is dense in Λ'' (since $L_0^\infty \subset \Lambda$) and hence $\widetilde{\Lambda} = \Lambda''$. If, on the contrary $\widetilde{w} \in L^1$, $\Lambda^1(\widetilde{w})_0$ is closed in $\Lambda^1(\widetilde{w})$ and it contains the space $\Lambda^p(w)$. If $f \in \Lambda^1(\widetilde{w})_0$, the functions $f_n = f\chi_{\{|f|>f^*(n)\}}$ have support in a set of finite measure and converge to f in $\Lambda^1(\widetilde{w})$. Each of these functions f_n has absolutely continuous norm (Corollary 2.3.5) and thus f_n is limit in $\Lambda^1(\widetilde{w})$ of function in L_0^∞. It follows that this space is dense in $\Lambda^1(\widetilde{w})_0$ and since $L_0^\infty \subset \Lambda^p(w)$, we conclude that the Banach envelope of $\Lambda^p(w)$ is $\Lambda^1(\widetilde{w})_0$. In the atomic case, $\widetilde{\Omega} \in \ell^1$ implies $\Lambda'' = d(\widetilde{\Omega}, 1) = \ell^\infty$ and $\widetilde{\Lambda} = c_0$. $\square$

Remark 2.4.29 Conditions (a) and (c) in (i) of the previous theorem are natural, since if $W \notin \Delta_2$ there is no topology in $\Lambda^p(w)$ and, if condition (c) fails, $\Lambda^* = \{0\}$ (Theorem 2.4.9) and it does not make sense to consider the Mackey topology in $\Lambda^p(w)$ with "respect to the dual pair $(\Lambda^p(w), \Lambda^p(w)^*)$". The conditions on Ω in (ii) are not at all restrictive, since if $\Omega \in \ell^1$, $d(\Omega, 1) = \ell^\infty$ and the Mackey completion is, in this case, the same space $d(\Omega, p) = \ell^\infty$.

N. Popa proves in [Po], for the case $\Omega \downarrow$, that the Mackey completion of $d(\Omega, p)$ ($0 < p < 1$) is ℓ^1 if and only if $d(\Omega, p) \subset \ell^1$. By Theorem 2.4.28 we can extend this result to the general case.

Corollary 2.4.30 *Let $0 < p \leq 1$, $\Omega \notin \ell^1$. Then the Banach envelope of $d(\Omega, p)$ is ℓ^1 if and only if $d(\Omega, p) \subset \ell^1$.*

Proof. If $d(\Omega, p) \subset \ell^1$ then $\ell^\infty = (\ell^1)' \subset d(\Omega, p)'$ and it follows (Theorem 2.4.17) that these last two spaces are equal, and hence $d(\Omega, p)'' = \ell^1$. Since $d(\Omega, p)'' = d(\widetilde{\Omega}, 1)$ for some sequence $\widetilde{\Omega}$ (Theorem 2.4.18), it follows that $\widetilde{\Omega} \notin \ell^1$ and, by the previous theorem, the Banach envelope of $d(\Omega, p)$ is $d(\Omega, p)'' = \ell^1$.

The converse is immediate. $\square$

Let us now study the weak-type spaces $\Lambda^{p,\infty}$. First we shall consider the nonatomic case X.

Theorem 2.4.31 *If $0 < p < \infty$, $W \in \Delta_2$ and X is nonatomic, then $\Lambda_X^{p,\infty}(w)' = \Lambda_X^{p,\infty}(w)^*$ if and only if these two spaces are zero.*

Proof. The sufficiency is obvious. To see the necessity, we only need to prove that $\Lambda_X^{p,\infty}(w)' \neq \Lambda_X^{p,\infty}(w)^*$ if the first of these spaces is not trivial.

But if $\Lambda_X^{p,\infty}(w)' \neq \{0\}$, the function $W^{-1/p}$ is locally integrable in $[0, \infty)$ (Theorem 2.4.9) and we can define the seminorm

$$H(f) = \limsup_{t \to 0} \frac{\int_0^t f^*(s)\,ds}{\int_0^t W^{-1/p}(s)\,ds}, \qquad f \in \Lambda^{p,\infty}.$$

If $f \in \Lambda_X^{p,\infty}(w)$ we have that $f^*(t) \leq \|f\|_{\Lambda^{p,\infty}} W^{-1/p}(t)$, and therefore H is well defined and it is continuous. Moreover, it is not zero because there exists $f_0 \in \Lambda^{p,\infty}$ with $f_0^* = W^{-1/p}$ in $[0, \mu(X))$ and hence $H(f_0) = 1$. By Hahn-Banach theorem, there exists a nonzero $u \in \Lambda^*$ such that $|u(f)| \leq H(f)$, for every f. This linear form is not in Λ', because it is zero over all functions $f \in L_0^\infty$:

$$H(f) = \limsup_{s \to 0} \frac{\int_0^s f^*(t)\,dt}{\int_0^s W^{-1/p}(t)\,dt} \leq \|f\|_{L^\infty} \limsup_{s \to 0} \frac{s}{\int_0^s W^{-1/p}(t)\,dt} = 0.$$

Thus $\Lambda_X^{p,\infty}(w)^* \neq \Lambda_X^{p,\infty}(w)'$. □

Theorem 2.4.32 *Let $0 < p < \infty$ and $d^\infty(\Omega, p)$ be quasi-normed. Let $W_n = \sum_{k=0}^n \Omega_k$, $n = 0, 1, \ldots$ Then, if $\Omega \notin \ell^1$ and $W^{-1/p} \notin \ell^1$, we have that $d^\infty(\Omega, p)^* \neq d^\infty(\Omega, p)'$.*

Proof. The seminorm

$$H(f) = \limsup_{n \to \infty} \frac{\sum_{k=0}^n f^*(k)}{\sum_{k=0}^n W_k^{-1/p}}, \qquad f \in d^\infty(\Omega, p),$$

is continuous and nonzero, and if $W^{-1/p} \notin \ell^1$, it is zero on all finite sequences. Thus, using and analogous argument as in the previous proof, there exists $u \in d^\infty(\Omega, p)^* \setminus d^\infty(\Omega, p)'$. □

In the nonatomic cases, it remains to characterize the identity $\Lambda_X^{1,\infty}(w)^* = \{0\}$. We first need the following results.

Lemma 2.4.33 *Let $W \in \Delta_2$. If $\mu(X) = \infty$ and $w \in L^1$, then $\Lambda^{1,\infty}(w)^* \neq \{0\}$.*

Proof. By hypothesis, we have that if $H(f) = \lim_{t \to \infty} f^*(t)$, then $H : \Lambda^{1,\infty}(w) \to [0, \infty)$ is a continuous seminorm in $\Lambda^{1,\infty}(w)$ not identically zero (since $1 \in \Lambda^{1,\infty}(w)$ and $H(1) = 1 \neq 0$). By Hahn-Banach theorem it follows that $\Lambda^{1,\infty}(w)^* \neq \{0\}$. □

Lemma 2.4.34 *Let X be nonatomic and $f \in \Lambda_X^{1,\infty}(w)$. If $\lim_{t \to \infty} f^*(t) = 0$ there exists $F \in \Lambda_X^{1,\infty}(w)$ satisfying:*

(i) $|f(x)| \leq F(x)$ a.e. $x \in X$.

(ii) $F^*(t) = \|f\|_{\Lambda^{1,\infty}} W^{-1}(t)$, $0 < t < \mu(X)$.

Proof. We can assume $\|f\|_{\Lambda^{1,\infty}} = 1$. Let $A = \{f \neq 0\}$, $a = \mu(A)$. Since $f^*(t) \xrightarrow[t \to \infty]{} 0$, there exists a measure preserving transformation $\sigma : A \to (0, a)$ such that $|f(x)| = f^*(\sigma(x))$ a.e. $x \in A$ (see Theorem II.7.6 in [BS]). Define

$$F_0(x) = W^{-1}(\sigma(x))\chi_A(x), \quad x \in X.$$

Since σ is measure preserving, $F_0^*(t) = W^{-1}(t)$, $0 < t < a$. Besides, since $f^* \leq W^{-1}$, we have that

$$F_0(x) \geq f^*(\sigma(x)) = |f(x)| \quad \text{a.e. } x \in A.$$

If $a = \mu(X)$ the function we are looking for is clearly, $F = F_0$. If on the contrary $a < \mu(X)$, we take $0 \leq F_1 \in \mathcal{M}(X \setminus A)$, with $F_1^*(t) = W^{-1}(a + t)$, $0 \leq t < \mu(X \setminus A)$. It is then immediate to check that the function $F = F_0 + F_1 \chi_{X \setminus A}$ satisfies the statement. □

Let us define now the seminorm N whose properties (Proposition 2.4.36) will be very useful.

Definition 2.4.35 *If $(E, \|\cdot\|)$ is a quasi-normed space, we define the seminorm $N : E \to [0, +\infty)$ by*

$$N(f) = N_E(f) = \inf\left\{\sum_{k=1}^K \|f_k\| : (f_k)_k \subset E, f = \sum_k f_k\right\}, \quad f \in E.$$

It can be proved that N is the bidual norm:

$$N(f) = \sup_{u \in \Lambda^*} \frac{|u(f)|}{\|u\|_{\Lambda^*}} = \|f\|_{\Lambda^{**}}.$$

Proposition 2.4.36 *Let Λ be a quasi-normed Lorentz space, and $N = N_\Lambda$. Then for $f, g \in \Lambda$ we have,*

(i) $N(f) \leq \|f\|_\Lambda$.

(ii) $|f| \leq |g| \Rightarrow N(f) \leq N(g)$.

(iii) $N(f) = N(|f|)$.

(iv) $N(f) = \inf \left\{ \sum_{k=1}^{K} \|f_k\|_\Lambda : |f| \leq \sum_k |f_k| \right\}$.

Moreover $\Lambda^* = \{0\}$ if and only if $N = 0$.

Proof. (i) is immediate. To see (ii), if $g = \sum_k g_k$ we have that $f = (f/g)g = \sum_k (f/g)g_k$, hence $N(f) \leq \sum_k \|(f/g)g_k\|_\Lambda \leq \sum_k \|g_k\|_\Lambda$ and thus $N(f) \leq N(g)$. (iii) and (iv) are corollaries of (ii).

Let us now prove the last statement. If there exists $0 \neq u \in \Lambda^*$, we can find $f \in \Lambda$ with $u(f) \neq 0$. For every finite decomposition $f = \sum_k f_k$ we then have that, $|u(f)| \leq \sum_k |u(f_k)| \leq \|u\| \sum_k \|f_k\|_\Lambda$. Hence $\sum_k \|f_k\|_\Lambda \geq |u(f)|/\|u\|$ and taking the infimum, we obtain $N(f) \geq |u(f)|/\|u\| > 0$, that is, $N \neq 0$. Conversely, since N is a continuous seminorm in Λ (by (i)), if $N \neq 0$ there exists (Hahn-Banach theorem, [Ru]) a nonzero continuous linear form u in Λ. □

In the following result, proved by A. Haaker in [Ha] for the case $X = \mathbb{R}^+$, we give a necessary condition to have that the dual of $\Lambda^{1,\infty}(w)$ is zero. As we shall see in Theorem 2.4.40, this condition is also sufficient. Our proof is based on the proof of A. Haaker.

Lemma 2.4.37 *Let X be nonatomic and $W \in \Delta_2$. If $\Lambda_X^{1,\infty}(w)^* = \{0\}$ and $\epsilon > 0$, there exists $n \in \mathbb{N}$ such that*

$$\int_{nt}^{s} W^{-1}(r)\,dr \leq \epsilon \int_{t}^{s} W^{-1}(r)\,dr, \quad 0 < t < nt < s \leq \mu(X).$$

Proof. We can assume that W is strictly increasing (otherwise, we substitute this function by $W(t)(1+2t)/(1+t)$). Let $0 \leq f \in \Lambda^{1,\infty}(w)$ with $f^* = W^{-1}$ in $(0, \mu(X))$. By Proposition 2.4.36, $\mathrm{N}(f) = 0$ and there exist positive functions $f_1, \ldots, f_{n-1} \in \Lambda^{1,\infty}(w)$ with

$$f \leq \sum_k f_k,$$

and such that, if $a_k = \|f_k\|_{\Lambda^{1,\infty}}$, $k = 1, \ldots, n-1$, then

$$\sum_k a_k < \epsilon.$$

By Lemma 2.4.33, $\mu(X) < \infty$ or $\mu(X) = \infty$ and $w \notin L^1$. In any case $\lim_{t \to \infty} f_k^*(t) = 0$, $k = 1, \ldots, n-1$. We can then take functions $g_1, \ldots, g_{n-1} \in \Lambda^{1,\infty}(w)$ with $g_k \geq f_k$ a.e. and $g_k^*(t) = a_k W^{-1}(t)$, $0 < t < \mu(X)$ (using Lemma 2.4.34). Summarizing,

$$
\begin{aligned}
(i) &\quad f \leq g_1 + \ldots + g_{n-1} \text{ a.e.,} \\
(ii) &\quad g_k^*(t) = a_k W^{-1}(t),\ 0 < t < \mu(X), \\
(iii) &\quad a_1 + \ldots + a_{n-1} < \epsilon.
\end{aligned}
$$

If $s, t > 0$, with $nt < s < \mu(X)$, we define

$$
\begin{aligned}
F &= \{x \in X : W^{-1}(s) < f(x) \leq W^{-1}(t)\}, \\
E_k &= \{x \in F : g_k(x) \leq a_k W^{-1}(t)\}, \quad k = 1, \ldots, n-1, \\
E &= \bigcap_{k=1}^{n-1} E_k.
\end{aligned}
$$

Since $f^* = W^{-1}$ and this function is strictly decreasing,

$$\mu(F) = \lambda_{f^*}(W^{-1}(s)) - \lambda_{f^*}(W^{-1}(t)) = s - t.$$

Moreover,

$$\mu(F \setminus E_k) = \mu(\{x \in F : g_k(x) > a_k W^{-1}(t)\}) \leq \lambda_{g_k^*}(a_k W^{-1}(t)) = t,$$

and $\mu(F \setminus E) = \mu((F \setminus E_1) \cup \ldots \cup (F \setminus E_{n-1})) \leq (n-1)t$. Hence,

$$\mu(E) \geq \mu(F) - (n-1)t = s - nt.$$

Since $E \subset F$, if $G = \{W^{-1}(s) < f \leq W^{-1}(s - \mu(E))\}$, the distribution function of $f\chi_E$ majorizes $f\chi_G$ and we have that

$$\int_E f(x)\,d\mu(x) = \int_0^\infty (f\chi_E)^*(r)\,dr \geq \int_0^\infty (f\chi_G)^*(r)\,dr \qquad (2.10)$$
$$= \int_{s-\mu(E)}^s W^{-1}(r)\,dr \geq \int_{nt}^s W^{-1}(r)\,dr.$$

On the other hand,

$$\int_E f(x)\,d\mu(x) \leq \sum_{k=1}^{n-1} \int_E g_k(x)\,d\mu(x)$$
$$\leq \sum_k \int_{E_k} g_k(x)\,d\mu(x) = \sum_k \int_0^\infty (g_k\chi_{E_k})^*(r)\,dr,$$

but $\mu(E_k) \leq \mu(F) = s - t$ and in this set $g_k \leq a_k W^{-1}(t) = g_k^*(t)$. Hence, if $H_k = \{g_k^*(t) \geq g_k > g_k^*(s)\}$, the distribution function of χ_{H_k} majorizes the distribution function of χ_{E_k} (observe that $g_k^* = a_k W^{-1}$ is strictly decreasing) and we have that

$$\int_0^\infty (g_k\chi_{E_k})^*(r)\,dr \leq \int_t^s g_k^*(r)\,dr = a_k \int_t^s W^{-1}(r)\,dr,$$

and hence,

$$\int_E f(x)\,d\mu(x) \leq \sum_k a_k \int_t^s W^{-1}(r)\,dr < \epsilon \int_t^s W^{-1}(r)\,dr.$$

This and (2.10) end the proof. $\square$

Definition 2.4.38 A function $f \in \mathcal{M}(X)$ is a step function if it is of the form

$$f = \sum_{n=1}^\infty a_n \chi_{E_n},$$

where $(a_n)_n \subset \mathbb{C}$ and $(E_n)_n$ is a sequence of pairwise disjoint measurable sets in X.

The proof of the following result is totally analogous to that of Lemma 7 in [Cw1], where the author considers the case $W(t) = t^{1/p}$.

Lemma 2.4.39 *Every function $0 \leq f \in \Lambda_X^{1,\infty}(w)$ is majorized by a step function $F \in \Lambda^{1,\infty}$.*

Now, we can prove the following result.

Theorem 2.4.40 *Let X be nonatomic and $W \in \Delta_2$. If $0 < p < \infty$,*

$$\Lambda_X^{p,\infty}(w)^* = \{0\}$$

if and only if

$$\lim_{t \to 0^+} \sup_{0 < r < \mu(X)} \frac{W^{1/p}(tr)}{tW^{1/p}(r)} = 0. \qquad (2.11)$$

Proof. By Remark 2.2.6, we can assume $p = 1$.

Sufficiency. We assume that (2.11) holds and we shall prove that $N = N_{\Lambda^{1,\infty}} = 0$. By Proposition 2.4.36 we have $\Lambda_X^{p,\infty}(w)^* = \{0\}$. Let then $f \in \Lambda^{1,\infty}$ and let us see that $N(f) = 0$. By Lemma 2.4.39 and Proposition 2.4.36 (ii), we can assume that f is a positive step function:

$$f = \sum_n a_n \chi_{A_n}, \quad (a_n)_n \subset (0,\infty), \ A_n \cap A_m = \emptyset, \ n \neq m.$$

Observe that either $\mu(X) < \infty$ or, by (2.11), $W(\infty) = \infty$. Therefore, $\lim_{t \to \infty} f^*(t) = 0$, and hence $\mu(A_n) < \infty$ for every $n = 1, 2, \ldots$ Also by (2.11), for every $\epsilon > 0$ there exists $m \in \mathbb{N}$ such that

$$2^m \frac{W(2^{-m} r)}{W(r)} < \epsilon, \quad 0 < r < \mu(X).$$

That is,
$$W^{-1}(2^m t) < \epsilon 2^{-m} W^{-1}(t), \quad 0 < t < 2^{-m} \mu(X).$$

Let us divide the sets A_n into 2^m pairwise disjoint measurable subsets $(A_n^k)_{k=1}^{2^m}$ and with equal measure. For every $k = 1, 2, \ldots, 2^m$, we have

$$f_k = \sum_n a_n \chi_{A_n^k}.$$

Thus $f = \sum_k f_k$ and for $0 < t < 2^{-m} \mu(X)$ (f_k^* is zero outside this interval) we have,

$$\begin{aligned} f_k^*(t) &= f^*(2^m t) \leq W^{-1}(2^m t) \|f\|_{\Lambda^{1,\infty}} \\ &< \epsilon 2^{-m} W^{-1}(t) \|f\|_{\Lambda^{1,\infty}}, \quad k = 1, \ldots, 2^m. \end{aligned}$$

Therefore, $\|f_k\|_{\Lambda^{1,\infty}} \leq \epsilon 2^{-m} \|f\|_{\Lambda^{1,\infty}}$ and,

$$N(f) \leq \sum_k \|f_k\|_{\Lambda^{1,\infty}} \leq \epsilon \|f\|_{\Lambda^{1,\infty}}.$$

Since this holds for every $\epsilon > 0$ we conclude that $N(f) = 0$.

Necessity. Let us assume now that $\Lambda_X^{1,\infty}(w)^* = \{0\}$. By Lemma 2.4.37, there exists $n \in \mathbb{N}$ such that

$$\int_{nt}^s W^{-1}(r)\,dr \leq \frac{1}{2} \int_t^s W^{-1}(r)\,dr, \quad t, s > 0,\ nt < s < \mu(X). \tag{2.12}$$

To see (2.11) we shall prove that for $K \in \mathbb{N}$,

$$\sup_{0 < r < \mu(X)} \frac{W(tr)}{tW(r)} \leq 4n\,2^{-K},$$

if $0 < t < n^{-2K}$. To this end, let us define for $0 < r < \mu(X)$

$$A_k = \int_{n^k tr}^r W^{-1}(x)\,dx, \quad k = 0, 1, \ldots, K.$$

By (2.12),
$$A_0 \geq 2A_1 \geq 4A_2 \geq \ldots \geq 2^K A_K$$

and
$$\frac{1}{2} \geq \frac{A_1}{A_0} = 1 - \frac{A_0 - A_1}{A_0} \geq 1 - 2^{-K}\frac{A_0 - A_1}{A_K}.$$

But $A_0 - A_1 = \int_{tr}^{ntr} W^{-1}(x)\,dx \leq (n-1)trW^{-1}(tr)$ and $A_K \geq (1 - n^K t)rW^{-1}(r)$. Hence,

$$\frac{1}{2} \geq 1 - 2^{-K}\frac{(n-1)tW(r)}{(1 - n^K t)W(rt)}.$$

And then,
$$\frac{W(tr)}{tW(r)} \leq 2^{1-K}\frac{n-1}{1 - n^K t} \leq 4n\,2^{-K}. \quad \square$$

Remark 2.4.41 (i) Condition (2.11) appears for the first time in [Ha] where A. Haaker proves the previous result for the case $X = \mathbb{R}^+$, $w \notin L^1$.

(ii) In the proof, we have seen that the condition of Lemma 2.4.37 implies (2.11). Therefore such condition is also equivalent to $\Lambda^{1,\infty}(w)^* = \{0\}$.

Corollary 2.4.42 *Let $0 < p < \infty$, X be nonatomic and $\Lambda = \Lambda_X^{p,\infty}(w)$ be quasi-normed. Then:*

(i) If $\lim_{t\to 0^+} \sup_{0<r<\mu(X)} \frac{W^{1/p}(tr)}{tW^{1/p}(r)} = 0$,

$$\{0\} = \Lambda' = \Lambda^*.$$

(ii) If the previous condition fails, two cases can occur:

(ii.1) If $\int_0^1 W^{-1/p}(s)\,ds = \infty$,

$$\{0\} = \Lambda' \subsetneq \Lambda^*.$$

(ii.2) If $\int_0^1 W^{-1/p}(s)\,ds < \infty$,

$$\{0\} \neq \Lambda' \subsetneq \Lambda^*.$$

Proof. It is an immediate consequence of Theorems 2.4.9, 2.4.31, and 2.4.40. □

Example 2.4.43

(i) We can apply the previous result to identify the dual and associate space of the Lorentz spaces $L^{p,q}(X)$ (X nonatomic), observing that $L^{p,q} = \Lambda^q(t^{q/p-1})$ and $L^{p,\infty} = \Lambda^{p,\infty}(1)$. We obtain, then, by other methods, results already known (see [Hu, CS, Cw1, Cw2]):

(i.1) $L^{p,q}(X)' = L^{p,q}(X)^* = \{0\}$, if $0 < p < 1$, $0 < q < \infty$,
(i.2) $L^{1,q}(X)' = L^{1,q}(X)^* = L^\infty(X)$, if $0 < q \leq 1$,
(i.3) $L^{1,q}(X)' = L^{1,q}(X)^* = \{0\}$, if $1 < q < \infty$,
(i.4) $L^{p,q}(X)' = L^{p,q}(X)^* = L^{p',\infty}(X)$, if $1 < p < \infty$, $0 < q \leq 1$,
(i.5) $L^{p,q}(X)' = L^{p,q}(X)^* = L^{p',q'}(X)$, if $1 < p < \infty$, $1 < q < \infty$,
(i.6) $L^{p,\infty}(X)' = L^{p,\infty}(X)^* = \{0\}$, if $0 < p < 1$,
(i.7) $\{0\} = L^{1,\infty}(X)' \subsetneq L^{1,\infty}(X)^*$,
(i.8) $\{0\} \neq L^{p,\infty}(X)' \subsetneq L^{p,\infty}(X)^*$, if $1 < p < \infty$.

(ii) If $w = \chi_{(0,1)}$, $\Lambda_{\mathbb{R}^+}^p(w) \supset \Lambda_{\mathbb{R}^+}^p(1) = L^p(\mathbb{R}^+)$. $\Lambda_{\mathbb{R}^+}^p(w)^* \neq \{0\}$ for every $p \in (0,\infty)$ while, as for L^p, $\Lambda_{\mathbb{R}^+}^p(w)' \neq \{0\}$, if and only if $p \geq 1$. In the weak-type case, $\Lambda_{\mathbb{R}^+}^{p,\infty}(w)^* \neq \{0\}$ for every $p \in (0,\infty)$ and the associate space is not zero only if $p > 1$.

2.5 Normability

In this section (X, μ) is an arbitrary σ-finite measure space. We shall study when $\Lambda_X^{p,q}(w)$ is a Banach space in the sense that there exists a norm in $\Lambda_X^{p,q}(w)$ equivalent to the functional $\|\cdot\|_{\Lambda_X^{p,q}(w)}$. In the nonatomic case, some results are already known: G.G. Lorentz ([Lo2], 1951) proved that, for $p \geq 1$, $\|\cdot\|_{\Lambda_{(0,l)}^p(w)}$ is a norm if and only if w is decreasing (that is, equal a.e. to a decreasing function) in $(0,l)$. A. Haaker ([Ha], 1970) characterizes the normability of $\Lambda_{\mathbb{R}^+}^{p,\infty}(w)$, $0 < p < \infty$ (see also [So]), and gives some partial results for $\Lambda_{\mathbb{R}^+}^p(w)$, $p < \infty$. E. Sawyer ([Sa], 1990) gives a necessary and sufficient condition to have that $\Lambda_X^p(w)$ is normed in the case $1 < p < \infty$, $X = \mathbb{R}^n$, while in ([CGS], 1996) the authors solve the case $0 < p < \infty$, $\mu(X) = \infty$, X nonatomic.

In the case $X = \mathbb{N}^*$ (spaces $d(\Omega, p)$) we do not know any previous result concerning normability. Except some isolated case, in all the references we know on the spaces $d(\Omega, p)$, it is assumed that the sequence Ω is decreasing which, in the case $p \geq 1$, ensures that $\|\cdot\|_{d(\Omega,p)}$ is already a norm.

Here, we solve the general case for a resonant measure space X unifying in this way all the previous results and including the unknown results for the spaces $d(\Omega, p)$. In the atomic case we shall reduce the problem to $\mathbb{R}^n$.

Let us start with some general results.

Theorem 2.5.1 *If $1 \leq p < \infty$ and $w \downarrow$, then $\|\cdot\|_{\Lambda_X^p(w)}$ is a norm.*

Proof. In the nonatomic case,

$$\|f + g\|_{\Lambda_X^p(w)} = \left(\int_0^\infty ((f+g)^*(t))^p w(t)\, dt\right)^{1/p}$$
$$= \sup_{h^* = w} \left(\int_X |f(x) + g(x)|^p h(x)\, d\mu(x)\right)^{1/p},$$

and the triangular inequality is immediate. On the other hand if X is σ-finite, we know (retract method) that there exists a nonatomic space $\bar{X}$ and a linear application $F : \mathcal{M}(X) \to \mathcal{M}(\bar{X})$ such that $F(f)^* = f^*$ (c.f. [BS]) and the result follows immediately from the atomic case. $\square$

The following result gives sufficient conditions to have the normability.

Theorem 2.5.2 *Let X be a σ-finite measure space. Then,*

(i) if $1 \leq p < \infty$ and $w \in B_{p,\infty}$, $\Lambda_X^p(w)$ is normable,

(ii) if $0 < p < \infty$ and $w \in B_p$, $\Lambda_X^{p,\infty}(w)$ is normable.

Proof. (i) If $p = 1$ and $w \in B_{p,\infty} = B_{1,\infty}$, there exists a decreasing function $\tilde{w}$ with $\widetilde{W} \approx W$ ([CGS]). It follows that $\|\cdot\|_{\Lambda^1(\tilde{w})}$ is a norm in $\Lambda^1(w)$ equivalent to the original one. If $p > 1$ and $w \in B_{p,\infty}$, the Hardy operator A satisfies $A: L^p_{\text{dec}}(w) \to L^p(w)$ and it follows that the functional $\|f\| = \|f^{**}\|_{L^p(w)}$ is an equivalent norm to the original quasi-norm.

(ii) If $w \in B_p$ we have that $A: L^{p,\infty}_{\text{dec}}(w) \to L^{p,\infty}(w)$ (see [So]) and the functional $\|f\| = \|f^{**}\|_{L^{p,\infty}(w)}$ is an equivalent norm. □

Remark 2.5.3 The functional $\|\cdot\|_{\Lambda^{p,\infty}}$ is not a norm, except in some trivial cases. In fact, if (X, μ) is an arbitrary measure space and there exist two measurable sets $E \subset F \subset X$ with $1 < a = W(\mu(F))/W(\mu(E)) < \infty$ (where w is an arbitrary weight in $\mathbb{R}^+$), we have that $\|f + g\|_{\Lambda^{p,\infty}} > \|f\|_{\Lambda^{p,\infty}} + \|g\|_{\Lambda^{p,\infty}}$ for $f = a\chi_E + \chi_{F \setminus E}$, $g = \chi_E + a\chi_{F \setminus E}$ (assuming without loss of generality that $\mu(F \setminus E) \leq \mu(E)$). Then, the following statements are equivalent:

(i) $\|\cdot\|_{\Lambda^{p,\infty}}$ is a norm.
(ii) $\|\cdot\|_{\Lambda^{p,\infty}} = C\|\cdot\|_{L^\infty}$.
(iii) The restriction of W to the range of μ is constant.

Theorem 2.5.4 *A Lorentz space Λ is normable if and only if $\Lambda = \Lambda''$, with equivalent norms. In particular, every normable Lorentz space Λ is a Banach function space with the norm $\|\cdot\|_{\Lambda''}$.*

Proof. The sufficiency is obvious. Conversely, if Λ is normable with a norm $\|\cdot\|$, it has to be equivalent to $\|\cdot\|_{\Lambda^{**}}$, (by Hahn-Banach theorem), and by Corollary 2.4.24, we have $\|f\|_\Lambda \approx \|f\|_{\Lambda''}$ for every function f with absolutely continuous norm in Λ. Since every $|f| \in \Lambda$ is a pointwise limit of an increasing sequence of functions in L_0^∞ (which have absolutely continuous norm by Corollary 2.3.5 and Proposition 2.3.8) and the functionals $\|\cdot\|_\Lambda$ and $\|\cdot\|_{\Lambda''}$ have the Fatou property, it follows that $\|f\|_\Lambda \approx \|f\|_{\Lambda''}$ for every $f \in \Lambda$. Finally, it is immediate to prove that the norm $\|\cdot\|_{\Lambda''}$ satisfies the properties of Definition 2.4.3. □

Remark 2.5.5 If $\|\cdot\|_\Lambda$ is a norm, then $(\Lambda, \|\cdot\|_\Lambda)$ is a Banach function space and $\|\cdot\|_\Lambda = \|\cdot\|_{\Lambda''}$ (see [BS]).

From now on, we shall consider the case X resonant. We shall see how the previous conditions are also necessary (in the atomic case). As an immediate consequence of the previous theorem and the representation theorem of Luxemburg ([BS]) we have the following result.

Corollary 2.5.6 *Let X be a nonatomic measure space, $0 < p, q \leq \infty$. If $\Lambda_X^{p,q}$ is normable, so is $\Lambda_{(0,\mu(X))}^{p,q}$, and if $\|\cdot\|_{\Lambda_X^{p,q}}$ is a norm the same holds for $\|\cdot\|_{\Lambda_{(0,\mu(X))}^{p,q}}$.*

Lemma 2.5.7 *Let $A \subset \mathbb{R}$ be a measurable set, and let w be a weight with $w = w\chi_{(0,|A|)}$. Then,*

(i) $\|\cdot\|_{\Lambda_A(w)}$ *is a norm if and only if* $\|\cdot\|_{\Lambda_\mathbb{R}(w)}$ *is a norm,*

(ii) $\Lambda_A(w)$ *is normable if and only if* $\Lambda_\mathbb{R}(w)$ *is normable.*

Proof. Since $\Lambda_A(w) \subset \Lambda_\mathbb{R}(w)$ the "if" implication is immediate. To see the other part, let us assume that

$$\Big\|\sum_{j=1}^n f_j\Big\|_{\Lambda_A(w)} \leq C \sum_j \|f_j\|_{\Lambda_A(w)}, \quad \forall f_1, \ldots, f_n \in \Lambda_A(w),$$

and let us first see that we have an analogous inequality substituting A by an arbitrary set $Y \subset \mathbb{R}$ with $|Y| \leq |A|$. By monotonicity we can assume $|Y| < |A|$. Then there exists an open set $G \supset Y$, with $b = |G| < |A|$. Let $\tilde{A} \subset A$ with $|\tilde{A}| = b$ and let $\sigma_1 : \tilde{A} \to (0, b)$ be a measure preserving transformation (Proposition 7.4 in [BS]). Since G is a countable union of intervals, it is easy to construct another measure preserving transformation $\sigma_2 : (0, b) \to G$. Then $\sigma = \sigma_2 \circ \sigma_1 : \tilde{A} \to G$ preserves the measure and for each $f_1, \ldots, f_n \in \mathcal{M}(Y) \subset \mathcal{M}(G)$, the functions $\tilde{f}_j = f_j \circ \sigma \in \mathcal{M}(A)$, $j = 1, \ldots, n$, satisfy $\tilde{f}_j^* = f_j^*$, for every j and $(\tilde{f}_1 + \ldots + \tilde{f}_n)^* = (f_1 + \ldots + f_n)^*$. Hence

$$\|f_1 + \ldots + f_n\|_{\Lambda_Y(w)} = \|\tilde{f}_1 + \ldots + \tilde{f}_n\|_{\Lambda_A(w)} \leq C \sum_j \|\tilde{f}_j\|_{\Lambda_A(w)} = C \sum_j \|f_j\|_{\Lambda_Y(w)}.$$

If now $f_1, \ldots, f_n$ are arbitrary measurable functions in $\mathbb{R}$ and we choose $Y \subset \{|f_1 + \ldots + f_n| \geq (f_1 + \ldots + f_n)^*(|A|)\}$, with $|Y| = |A|$, since w is supported in $(0, |A|)$ we have that

$$\|f_1 + \ldots + f_n\|_{\Lambda_\mathbb{R}(w)} = \|(f_1 + \ldots + f_n)\chi_Y\|_{\Lambda_\mathbb{R}(w)} = \Big\|\sum_j f_j \chi_Y\Big\|_{\Lambda_Y(w)},$$

and it follows that

$$\|f_1 + \ldots + f_n\|_{\Lambda_{\mathbb{R}}(w)} \leq C \sum_j \|f_j \chi_Y\|_{\Lambda_Y(w)} \leq C \sum_j \|f_j\|_{\Lambda_{\mathbb{R}}(w)}.$$

If $C = 1$ we conclude that $\|\cdot\|_{\Lambda_{\mathbb{R}}(w)}$ is a norm. If $1 < C < \infty$, the previous inequality proves that $\Lambda_{\mathbb{R}}(w)$ is normable (the functional N defined in 2.4.35 would be, for example, an equivalent norm). □

The following result characterizes the normability in the nonatomic case.

Theorem 2.5.8 *Let (X, μ) be a nonatomic measure space, $0 < p < \infty$, $w = w\chi_{(0, \mu(X))}$. Then:*

(i) $\|\cdot\|_{\Lambda_X^p(w)}$ *is a norm if and only if $p \geq 1$, $w \downarrow$.*

(ii) $\Lambda_X^p(w)$ *is normable if and only if $p \geq 1$, $w \in B_{p,\infty}$.*

(iii) $\|\cdot\|_{\Lambda_X^{p,\infty}(w)}$ *is not a norm (except if $\mu(X) = 0$).*

(iv) $\Lambda_X^{p,\infty}(w)$ *is normable if and only if $w \in B_p$.*

Proof. That the conditions are sufficient has been already seen in Theorems 2.5.1, 2.5.2, and Remark 2.5.3. To see the necessity, we use Corollary 2.5.6 and Lemma 2.5.7 to conclude that we can substitute the space X by $\mathbb{R}$. Then (i) follows from [CGS] and [Lo2]; (ii) from [CGS] and [Sa]; (iii) from Remark 2.5.3, and (iv) follows from [So]. □

We shall now study the normability of the sequence Lorentz spaces $d(\Omega, p)$ and $d^\infty(\Omega, p)$.

Theorem 2.5.9 *Let $\Omega = (\Omega_n)_{n=0}^\infty \subset [0, \infty)$.*

(i) *If $0 < p < 1$, $\|\cdot\|_{d(\Omega,p)}$ is a norm if and only if $\Omega = (\Omega_0, 0, 0, \ldots)$.*

(ii) *If $1 \leq p < \infty$, $\|\cdot\|_{d(\Omega,p)}$ is a norm if and only if $\Omega \downarrow$.*

Proof. The sufficiency in (i) is obvious and in (ii) has already been proved in Theorem 2.5.1. Let us see that if $\|\cdot\|_{d(\Omega,p)}$ is a norm, necessarily $\Omega \downarrow$.

For $n \in \mathbb{N}$ and $t \in (0,1)$, let $f = (1,1,\ldots,1,t,0,0,0,\ldots)$ and let $g = (1,\ldots,1,t,1,0,0\ldots)$. Then,

$$\|f\|_{d(\Omega,p)} = \|g\|_{d(\Omega,p)} = (\Omega_0 + \ldots + \Omega_n + t^p\Omega_{n+1})^{1/p},$$
$$\|f+g\|_{d(\Omega,p)} = (2^p\Omega_0 + \ldots + 2^p\Omega_{n-1} + (1+t)^p(\Omega_n + \Omega_{n+1}))^{1/p}.$$

Form the inequality $\|f+g\|_{d(\Omega,p)} \leq \|f\|_{d(\Omega,p)} + \|g\|_{d(\Omega,p)}$ it follows that

$$\Omega_{n+1} \leq \frac{2^p - (1+t)^p}{(1+t)^p - 2^p t^p} \Omega_n,$$

and letting t tend to 1 we obtain that $\Omega_{n+1} \leq \Omega_n$ and (ii) is proved.

Let us assume now that $\|\cdot\|_{d(\Omega,p)}$ is a norm and $p < 1$. Then (Remark 2.5.5) it coincides with the norm of the biassociate space. Then since by Theorem 2.4.14 (i), $d(\Omega,p)' = d(\widetilde{\Omega},1)'$, where $\widetilde{\Omega}_0 = \Omega_0^{1/p}$, $\widetilde{\Omega}_n = (\sum_{k=0}^n \Omega_k)^{1/p} - (\sum_{k=0}^{n-1} \Omega_k)^{1/p}$, $n = 1, 2, \ldots$, we obtain that $\|\cdot\|_{d(\Omega,p)''} = \|\cdot\|_{d(\widetilde{\Omega},1)''}$. Applying this equality of norms to $f = (1,t,0,0,\ldots)$, $t \in (0,1)$, we obtain that

$$(\Omega_0 + t^p\Omega_1)^{1/p} = \|f\|_{d(\Omega,p)} = \|f\|_{d(\widetilde{\Omega},1)''} \leq \|f\|_{d(\widetilde{\Omega},1)}$$
$$= \Omega_0^{1/p} + t((\Omega_0 + \Omega_1)^{1/p} - \Omega_0^{1/p}),$$

and hence

$$\frac{(\Omega_0 + t^p\Omega_1)^{1/p} - \Omega_o^{1/p}}{t} \leq C < \infty.$$

If $\Omega_1 \neq 0$ the limit, when $t \to 0$, of the left hand side is $+\infty$, and thus $\Omega_1 = 0$ and, since the sequence is decreasing, we finally obtain $\Omega_n = 0$, $n \geq 1$. □

Let us consider now the normability of $d(\Omega,p)$.

Theorem 2.5.10 Let $\Omega = (\Omega_n)_{n=0}^\infty \subset [0,\infty)$ and let us denote $W_n = \sum_{k=0}^n \Omega_k$, $n = 0, 1, 2, \ldots$

(i) If $0 < p < 1$, $d(\Omega,p)$ is normable if and only if $\Omega \in \ell^1$.

(ii) $d(\Omega,1)$ is normable if and only if

$$\frac{W_n}{n+1} \leq C \frac{W_m}{m+1}, \quad 0 < m < n.$$

(iii) If $1 < p < \infty$, $d(\Omega, p)$ is normable if and only if

$$\sum_{k=0}^{n} \frac{1}{W_k^{1/p}} \leq C \frac{n+1}{W_n^{1/p}}, \qquad n = 0, 1, 2, \ldots$$

That is, if $p < 1$, $d(\Omega, p)$ is not normable except in the trivial case $\Omega \in \ell^1$ and then $d(\Omega, p) = \ell^\infty$.

Proof. (i) If $\Omega \in \ell^1$, $d(\Omega, p) = \ell^\infty$ and there is nothing to prove. Conversely, if $d(\Omega, p)$ is normable, $\|\cdot\|_{d(\Omega,p)''}$ is a norm in this space (Theorem 2.5.4) which is majorized by the norm of $d(\widetilde{\Omega}, 1)$, with $\widetilde{\Omega}_0 = \Omega_0^{1/p}$, $\widetilde{\Omega}_n = \left(\sum_{k=0}^{n} \Omega_k\right)^{1/p} - \left(\sum_{k=0}^{n-1} \Omega_k\right)^{1/p}$, $n = 1, 2, \ldots$ (see the proof of the previous theorem). We have then the inequality,

$$\left(\sum_{n=0}^{\infty} g(n)^p \Omega_n\right)^{1/p} \leq C \sum_{n=0}^{\infty} g(n) \widetilde{\Omega}_n, \qquad g \downarrow .$$

If $r = 1/p > 1$ the previous expression is equivalent to

$$S = \sup_{g\downarrow} \frac{\sum_{n=0}^{\infty} g(n) \Omega_n}{\left(\sum_{n=0}^{\infty} g(n)^r \widetilde{\Omega}_n\right)^{1/r}} < \infty.$$

By Theorem 1.3.6,

$$S \approx \left(\int_0^\infty \left(\frac{W(t)}{\widetilde{W}(t)}\right)^{p/(1-p)} w(t)\, dt\right)^{1-p},$$

with $w = \sum_{k=0}^{\infty} \Omega_k \chi_{[k,k+1)}$, $W(t) = \int_0^t w(s)\, ds$ and analogously $\tilde{w}$ and $\widetilde{W}$. Using that $W \in \Delta_2$ (because $d(\Omega, p)$ is quasi-normed), one can easily see that the above integrand is comparable, in every interval $[n, n+1)$, to the function w/W. We have then

$$\int_1^\infty \frac{w(t)}{W(t)}\, dt = \log \frac{W(\infty)}{W(1)} \lesssim S < \infty,$$

which implies $w \in L^1$ and $\Omega \in \ell^1$.

(ii) and (iii): Let $w(t) = \sum_{k=0}^{\infty} \Omega_k \chi_{[k,k+1)}(t)$, $t > 0$. Then the condition of the statement implies $w \in B_{p,\infty}$ (c.f. Theorem 1.3.3 and Corollary 1.3.9) and by Theorem 2.5.2, $d(\Omega, p) = \Lambda_{\mathbb{N}^*}^p(w)$ is normable. To see the converse,

let $f \geq 0$ be a decreasing sequence in $d(\Omega, p)$. If $n \geq 1$, let us define $g_n = \sum_{j=-n}^{n} f_j$ where, for each $j \in \mathbb{N}^*$, $f_j(k) = f(k+j)\chi_{\{k \geq -j\}}(k)$, $k = 0, 1, 2, \ldots$
Then

$$\|g_n\|_{d(\Omega,p)} \leq C \sum_{j=-n}^{n} \|f_j\|_{d(\Omega,p)} \leq (2n+1)C\|f\|_{d(\Omega,p)}, \quad \forall n, \qquad (2.13)$$

and on the other hand, we have $g_n \geq (f(0) + \ldots + f(n))\chi_{\{0,\ldots,n\}}$ and hence $\|g_n\|_{d(\Omega,p)} \geq (f(0) + \ldots + f(n))\|\chi_{\{0,\ldots,n\}}\|_{d(\Omega,p)} = (f(0) + \ldots + f(n))W_n^{1/p}$. Combining this with (2.13) we obtain that

$$A_d f(n) W_n^{1/p} \leq 2C\|f\|_{d(\Omega,p)} = 2C\|f\|_{\ell^p(\Omega)}, \quad n \in \mathbb{N}, \; f \downarrow,$$

which is equivalent to the boundedness of $A_d : \ell^p_{\mathrm{dec}}(\Omega) \to \ell^{p,\infty}(\Omega)$. Now, by Theorem 1.3.7, the conditions of the statement follow. □

To finish this section, we characterize the normability in the weak case.

Theorem 2.5.11 *Let $0 < p < \infty$, $\Omega = (\Omega_n)_{n=0}^{\infty} \subset [0, \infty)$ and let us denote by $W_n = \sum_{k=0}^{n} \Omega_k$, $n = 0, 1, 2, \ldots$ Then:*

(i) *$\|\cdot\|_{d^\infty(\Omega,p)}$ is a norm if and only if $\Omega = (\Omega_0, 0, 0, \ldots)$.*

(ii) *$d^\infty(\Omega, p)$ is normable if and only if*

$$\sum_{k=0}^{n} \frac{1}{W_k^{1/p}} \leq C \frac{n+1}{W_n^{1/p}}, \quad n = 0, 1, 2, \ldots$$

Proof. (i) is a consequence of Remark 2.5.3. To see (ii) it is enough to observe that the normability of $d^\infty(\Omega, p)$ is equivalent to the boundedness of

$$A_d : \ell^{p,\infty}_{\mathrm{dec}}(\Omega) \to \ell^{p,\infty}(\Omega), \qquad (2.14)$$

since if (2.14) holds, then $\|f\| = \|A_d f^*\|_{\ell^{p,\infty}(\Omega)}$ is a norm in $d^\infty(\Omega, p)$ equivalent to the original quasi-norm, and if $d^\infty(\Omega, p)$ is normable, the same argument used in the last part of the previous theorem (changing $\|f\|_{d(\Omega,p)}$ by $\|f\|_{d^\infty(\Omega,p)}$) shows (2.14). Applying now Theorem 1.3.8 we conclude the proof. □

2.6 Interpolation of operators

In this section (X, μ) and $(\bar{X}, \bar{\mu})$ are σ-finite measure spaces, although this will not be necessary in the interpolation theorems for the spaces $\Lambda_X^{p,q}(w)$. In some cases (Theorem 2.6.3) no conditions on the weight w will be required. In the more general result (Theorem 2.6.5) we will assume that the spaces involved are quasi-normed ($W \in \Delta_2(X)$). This theorem is a generalization of Marcinkiewicz theorem adapted to the context of the $\Lambda^{p,q}(w)$ spaces.

Finally, we shall see how some of the results of chapter 1 on boundedness of order continuous operators can be extended to the context of Lorentz spaces.

We start by recalling some concepts connected with the real method of interpolation. In this section we call functional lattice to any class A formed by measurable functions and defined by

$$A = \{f : \|f\|_A < \infty\},$$

where $\|\cdot\|_A$ is a nonnegative functional that acts on measurable functions and satisfies $\|rf\|_A = r\|f\|_A$, $r > 0$, and $\|f\|_A \leq \|g\|_A$ if $|f(x)| \leq |g(x)|$ a.e. x. In particular the Lorentz spaces $\Lambda_X^{p,q}(w)$ are of this type. In what follows A, B, A_0, A_1, B_0, B_1 denote arbitrary functional lattices. We shall write $A \approx B$, if $A = B$ and $\|\cdot\|_A \approx \|\cdot\|_B$. Let us now recall the definition of the K-functional associated to the pair (A_0, A_1). For every function $f \in A_0 + A_1$ and $t > 0$

$$K(f, t, A_0, A_1) = \inf_{f=f_0+f_1} (\|f_0\|_{A_0} + t\|f_1\|_{A_1}),$$

where the infimum extends over all possible decompositions $f = f_0 + f_1$, $f_i \in A_i$, $i = 0, 1$. We shall simply write $K(t, f)$ if it is clear which pair (A_0, A_1) we are working with. The main properties of K are the following:

(i) $K(rf, t) = rK(f, t)$, $r > 0$,
(ii) $|f| \leq |g| \Rightarrow K(f, t) \leq K(g, t)$,
(iii) $A_i \approx B_i$, $i = 0, 1 \Rightarrow K(f, t, A_0, A_1) \approx K(g, t, B_0, B_1)$.

Properties (i) and (iii) are immediate while (ii) can be seen in [KPS]. For each $0 < \theta < 1$, $0 < q \leq \infty$, we define the following "norm" in $A_0 + A_1$:

$$\|f\|_{(A_0,A_1)_{\theta,q}} = \left(\int_0^\infty (t^{-\theta} K(f, t, A_0, A_1))^q \frac{dt}{t} \right)^{1/q},$$

($\sup_t t^{-\theta} K(f, t, A_0, A_1)$ if $q = \infty$). It is then natural to define the space

$$(A_0, A_1)_{\theta,q} = \{f \in A_0 + A_1 : \|f\|_{(A_0,A_1)_{\theta,q}} < \infty\}.$$

We say that the operator T, defined in $A_0 + A_1$, and with values in $B_0 + B_1$, is quasi-additive if there exists $k > 0$ such that

$$|T(f+g)(x)| \leq k(|Tf(x)| + |Tg(x)|) \quad \text{a.e. } x,$$

for every pair of functions $f, g, f+g \in A_0 + A_1$.

The fundamental result of this theory is the following.

Theorem 2.6.1 *Let T be a quasi-additive operator defined in $A_0 + A_1$ and such that*

$$\begin{aligned} T &: A_0 \longrightarrow B_0, \\ T &: A_1 \longrightarrow B_1. \end{aligned}$$

Then, for $0 < \theta < 1$, $0 < q \leq \infty$, we have,

$$T : (A_0, A_1)_{\theta,q} \longrightarrow (B_0, B_1)_{\theta,q}.$$

Our purpose now is to identify the space $(A_0, A_1)_{\theta,q}$ or, equivalently, the functional $\|\cdot\|_{(A_0,A_1)_{\theta,q}}$ when A_0, A_1 are Lorentz spaces. As we shall see, $\|\cdot\|_{(A_0,A_1)_{\theta,q}}$ will be, under appropriate conditions, equivalent to the "norm" of a certain Lorentz space.

Theorem 2.6.2 *If $0 < p < \infty$ and $f \in \mathcal{M}(X)$,*

$$K(f, t, \Lambda_X^p(w), L^\infty(X)) \approx K(f^*, t, L^p(w), L^\infty(w)), \qquad t > 0,$$

with constants depending only on p. In particular, for $0 < \theta < 1$, $0 < q \leq \infty$,

$$(\Lambda_X^p(w), L^\infty(X))_{\theta,q} \approx \Lambda_X^{\bar{p},q}(w)$$

where,

$$\frac{1}{\bar{p}} = \frac{1-\theta}{p}.$$

Proof. If $f = f_0 + f_1$ with $f_0 \in \Lambda_X^p(w)$, $f_1 \in L^\infty(X)$, we have that $f^*(s) \leq f_0^*(s) + f_1^*(0) = f_0^*(s) + \|f_1\|_{L^\infty(X)}$, $s > 0$. Hence,

$$K(f^*, t, L^p(w), L^\infty(w)) \leq \|f_0^*\|_{L^p(w)} + t\|f_1\|_{L^\infty(X)} = \|f_0\|_{\Lambda_X^p(w)} + t\|f_1\|_{L^\infty(X)}.$$

Taking the infimum over all decompositions $f = f_0 + f_1 \in \Lambda_X^p(w) + L^\infty(X)$ we obtain,

$$K(f^*, t, L^p(w), L^\infty(w)) \le K(f, t, \Lambda_X^p(w), L^\infty(X)).$$

To prove the converse inequality, if $f \in \mathcal{M}(X)$, $t > 0$ let $a = (f^*)_w^*(t^p)$ and let

$$f_0 = \left(f - a\frac{f}{|f|}\right)\chi_{\{|f|>a\}}, \qquad f_1 = f - f_0.$$

Then $(f_0^*)_w^* = ((f^*)_w^* - a)\chi_{[0,t^p)}$ while $f_1^* \le a$. Since $f = f_0 + f_1$ we have that

$$\begin{aligned}
K(f, t, \Lambda_X^p(w), L^\infty(X)) &\le \|f_0\|_{\Lambda_X^p(w)} + t\|f_1\|_{L^\infty(X)} \\
&\le \|f_0^*\|_{L^p(w)} + ta \\
&= \|(f_0^*)_w^*\|_p + ta \\
&= \left(\int_0^{t^p} ((f^*)_w^*(s) - a)^p \, ds\right)^{1/p} + \left(\int_0^{t^p} a^p \, ds\right)^{1/p} \\
&\le C_p\left(\int_0^{t^p} ((f^*)_w^*(s))^p \, ds\right)^{1/p}.
\end{aligned}$$

Since the last expression is equivalent to $K(f^*, t, L^p(w), L^\infty(w))$ (see [BL]), the first part of the theorem is proved.

The second part is an immediate consequence of the previous one. To see this, observe that the "norm" in $(\Lambda_X^p(w), L^\infty(X))_{\theta,q}$ is defined using the K-functional and we have

$$\|f\|_{(\Lambda_X^p(w), L^\infty(X))_{\theta,q}} = \|f^*\|_{(L^p(w), L^\infty(w))_{\theta,q}},$$

and the last "norm" is (see [BL]),

$$\|f^*\|_{L^{\bar{p},q}(w)} = \|f\|_{\Lambda_X^{\bar{p},q}(w)}. \qquad \square$$

A consequence of this result is the following interpolation theorem with $L^\infty(X)$.

Theorem 2.6.3 *If $0 < p, \bar{p} < \infty$ and T is a quasi-additive operator in $\Lambda_X^p(w) + L^\infty(X)$ such that,*

$$\begin{aligned}
T &: L^\infty(X) \longrightarrow L^\infty(X), \\
T &: \Lambda_X^p(w) \longrightarrow \Lambda_{\bar{X}}^{\bar{p},\infty}(\bar{w}),
\end{aligned}$$

then for $q, \bar{q} \in (0, \infty)$ satisfying $q/p = \bar{q}/\bar{p} > 1$, we have

$$T : \Lambda_X^{q,r}(w) \longrightarrow \Lambda_{\bar{X}}^{\bar{q},r}(\bar{w}), \qquad 0 < r \leq \infty.$$

Proof. The argument used in the first part of the previous theorem to prove the inequality

$$K(f^*, t, L^p(w), L^\infty(w)) \leq K(f, t, \Lambda_X^p(w), L^\infty(X)),$$

still works if we substitute the spaces $L^p(w)$ and $\Lambda_X^p(w)$ by $L^{\bar{p},\infty}(\bar{w})$ and $\Lambda_{\bar{X}}^{\bar{p},\infty}(\bar{w})$ respectively. Therefore, with $1/\bar{q} = (1-\theta)/\bar{p}$, it follows that (see [BL]),

$$\|Tf\|_{\Lambda_{\bar{X}}^{\bar{q},r}(\bar{w})} = \|(Tf)^*\|_{L^{\bar{q},r}(\bar{w})} = \|(Tf)^*\|_{\left(L^{\bar{p},\infty}(\bar{w}), L^\infty(w)\right)_{\theta,r}}$$
$$\leq \|Tf\|_{\left(\Lambda^{\bar{p},\infty}(\bar{w}), L^\infty\right)_{\theta,r}}$$

and the rest is a consequence of Theorem 2.6.1 and Theorem 2.6.2. □

Remark 2.6.4 Observe that in the last two results, it is not necessary that the Lorentz spaces involved are quasi-normed. If we assume that these spaces are quasi-normed (that is $W \in \Delta_2$) we can use the reiteration theorem to obtain a more general result analogous to the interpolation theorem of Marcinkiewicz.

Theorem 2.6.5 *Let $0 < p_i, q_i, \bar{p}_i, \bar{q}_i \leq \infty$, $i = 0, 1$, with $p_0 \neq p_1$, $\bar{p}_0 \neq \bar{p}_1$ and let T be a quasi-additive operator defined in $\Lambda_X^{p_0,q_0}(w) + \Lambda_X^{p_1,q_1}(w)$ satisfying*

$$T : \Lambda_X^{p_0,q_0}(w) \longrightarrow \Lambda_{\bar{X}}^{\bar{p}_0,\bar{q}_0}(\bar{w}),$$
$$T : \Lambda_X^{p_1,q_1}(w) \longrightarrow \Lambda_{\bar{X}}^{\bar{p}_1,\bar{q}_1}(\bar{w}).$$

Assume that $W \in \Delta_2(X)$ and $\bar{W} \in \Delta_2(\bar{X})$. Then, for $0 < \theta < 1$, $0 < r \leq \infty$,

$$T : \Lambda_X^{p,r}(w) \longrightarrow \Lambda_{\bar{X}}^{\bar{p},r}(\bar{w}),$$

where

$$\frac{1}{p} = \frac{1-\theta}{p_0} + \frac{\theta}{p_1}, \qquad \frac{1}{\bar{p}} = \frac{1-\theta}{\bar{p}_0} + \frac{\theta}{\bar{p}_1}.$$

Proof. Let $0 < s < \min\{p_0, p_1\}$ and take $\theta_0, \theta_1 \in (0,1)$ such that $1/p_i = (1-\theta_i)/s$, $i = 0, 1$. Then, by Theorem 2.6.2 it follows,

$$(\Lambda_X^{p_0,q_0}(w), \Lambda_X^{p_1,q_1}(w))_{\theta,r} \approx \left((\Lambda_X^s(w), L^\infty(X))_{\theta_0,q_0}, (\Lambda_X^s(w), L^\infty(X))_{\theta_1,q_1}\right)_{\theta,r}.$$

Since $W \in \Delta_2(X)$ we have that $\Lambda_X^s(w)$ is a quasi-Banach space and we can apply the reiteration theorem (Theorem 3.11.5 in [BL]) to identify the above space:

$$(\Lambda_X^s(w), \Lambda_X^\infty(w))_{\eta,r}$$

with $\eta = (1-\theta)\theta_0 + \theta\theta_1$. Applying again Theorem 2.6.2, we finally obtain

$$(\Lambda_X^{p_0,q_0}(w), \Lambda_X^{p_1,q_1}(w))_{\theta,r} \approx \Lambda_X^{p,r}(w)$$

since $(1-\eta)/s = (1-\theta)/p_0 + \theta/p_1$. The same argument works for $\Lambda_{\bar{X}}^{\bar{p}_i,\bar{q}_i}(\bar{w})$ and analogously,

$$(\Lambda_{\bar{X}}^{\bar{p}_0,\bar{q}_0}(\bar{w}), \Lambda_{\bar{X}}^{\bar{p}_1,\bar{q}_1}(\bar{w}))_{\theta,r} \approx \Lambda_{\bar{X}}^{\bar{p},r}(\bar{w}).$$

The result now follows from Theorem 2.6.1. □

Remark 2.6.6 J. Cerdà and J. Martín ([CM]) have proved, under some conditions on the weights w_0, w_1 that,

$$K(f, t, \Lambda^{p_0,r_0}(w_0), \Lambda^{p_1,r_1}(w_1)) \approx K(f^*, t, L^{p_0,r_0}(w_0), L^{p_1,r_1}(w_1))$$

for $0 < p_0, p_1, r_0, r_1 \leq \infty$. This allows them to obtain a more general interpolation theorem including the case

$$T : (\Lambda_X^{p_0,q_0}(w_0), \Lambda_X^{p_1,q_1}(w_1)) \longrightarrow (\Lambda_{\bar{X}}^{\bar{p}_0,\bar{q}_0}(\bar{w}_0), \Lambda_{\bar{X}}^{\bar{p}_1,\bar{q}_1}(\bar{w}_1)),$$

with $w_0 \neq w_1, \bar{w}_0 \neq \bar{w}_1$.

The following Marcinkiewicz interpolation type result is a direct consequence of Theorem 2.6.5.

Corollary 2.6.7 *Let $0 < p_0 < p_1 \leq \infty$, $W \in \Delta_2(X)$, and $\bar{W} \in \Delta_2(\bar{X})$. If T is a quasi-additive operator in $\Lambda_X^{p_0}(w) + \Lambda_X^{p_1}(w)$ such that,*

$$T : \Lambda_X^{p_0}(w) \longrightarrow \Lambda_{\bar{X}}^{p_0,\infty}(\bar{w}),$$
$$T : \Lambda_X^{p_1}(w) \longrightarrow \Lambda_{\bar{X}}^{p_1,\infty}(\bar{w}),$$

then, for $p_0 < p < p_1$,

$$T : \Lambda_X^p(w) \longrightarrow \Lambda_{\bar{X}}^p(\bar{w}).$$

Before ending this chapter, we shall extend some of the results of section 1.2 about boundedness of order continuous operators. The first result is a generalization of Corollary 1.2.12.

Theorem 2.6.8 *Let $L \subset \mathcal{M}(X)$ be a regular class and let $T: L \to \mathcal{M}(\bar{X})$ be a sublinear order continuous operator. If $0 < p_0 \leq 1$, $p_0 \leq p_1 < \infty$ and $w_1 \in B_{p_1/p_0,\infty}$, we have*

$$\|Tf\|_{\Lambda_{\bar{X}}^{p_1}(w_1)} \leq C\|f\|_{\Lambda_X^{p_0}(w_0)}, \qquad f \in L, \qquad (2.15)$$

if and only if there exists $C_0 < \infty$ such that

$$\|T\chi_B\|_{\Lambda_{\bar{X}}^{p_1}(w_1)} \leq C_0\|\chi_B\|_{\Lambda_X^{p_0}(w_0)}, \qquad \chi_B \in L.$$

Proof. We shall assume the last inequality and we shall prove (2.15). By monotonicity, it is sufficient to prove it for a simple function $f \geq 0$. Proceeding as in the proof of Theorem 1.2.11,

$$\|Tf\|_{\Lambda_{\bar{X}}^{p_1}(w_1)}^{p_0} = \||Tf|^{p_0}\|_{\Lambda_{\bar{X}}^{p_1/p_0}(w_1)} \leq \left\| \int_0^\infty p_0 t^{p_0-1} |T\chi_{\{f>t\}}(\cdot)|^{p_0}\, dt \right\|_{\Lambda_{\bar{X}}^{p_1/p_0}(w_1)}.$$

But the condition $w_1 \in B_{p_1/p_0,\infty}$ implies that $\|\cdot\|_{\Lambda_{\bar{X}}^{p_1/p_0}(w_1)}$ is equivalent to a Banach function norm (Theorem 2.5.2) and it follows,

$$\begin{aligned}
\|Tf\|_{\Lambda_{\bar{X}}^{p_1}(w_1)}^{p_0} &\leq C_1 \int_0^\infty p_0 t^{p_0-1} \||T\chi_{\{f>t\}}|^{p_0}\|_{\Lambda_{\bar{X}}^{p_1/p_0}(w_1)}\, dt \\
&= C_1 \int_0^\infty p_0 t^{p_0-1} \|T\chi_{\{f>t\}}\|_{\Lambda_{\bar{X}}^{p_1}(w_1)}^{p_0}\, dt \\
&\leq C_1 C_0^{p_0} \int_0^\infty p_0 t^{p_0-1} \|\chi_{\{f>t\}}\|_{\Lambda_X^{p_0}(w_0)}^{p_0}\, dt \\
&= C^{p_0} \int_0^\infty p_0 t^{p_0-1} W_0(\lambda_f(t))\, dt \\
&= C^{p_0} \|f\|_{\Lambda_X^{p_0}(w_0)}^{p_0}. \qquad \square
\end{aligned}$$

Using the same idea, one can easily prove an analogous result for the weak-type case:

Theorem 2.6.9 *Let $L \subset \mathcal{M}(X)$ be a regular class and let $T : L \to \mathcal{M}(\bar{X})$ be a sublinear order continuous operator. If $0 < p_0 \leq 1$, $0 < p_1 < \infty$ and $w_1 \in B_{p_1/p_0}$, we have*

$$\|Tf\|_{\Lambda_{\bar{X}}^{p_1,\infty}(w_1)} \leq C\|f\|_{\Lambda_X^{p_0}(w_0)}, \qquad f \in L,$$

if and only if, there exists $C_0 < \infty$ such that

$$\|T\chi_B\|_{\Lambda_{\bar{X}}^{p_1,\infty}(w_1)} \leq C_0\|\chi_B\|_{\Lambda_X^{p_0}(w_0)}, \qquad \chi_B \in L.$$

Combining the previous results with the general interpolation theorem (Theorem 2.6.5) we obtain a generalization of Stein and Weiss theorem on restricted weak-type operators ([SW]).

Theorem 2.6.10 *Let $0 < p_0, p_1, q_0, q_1 \leq \infty$, $p_0 \neq p_1$, $q_0 \neq q_1$ and let us assume that $T : (\Lambda_X^{p_0}(w) + \Lambda_X^{p_1}(w)) \to \mathcal{M}(\bar{X})$ is a sublinear order continuous operator satisfying*

$$\|T\chi_B\|_{\Lambda^{q_0,\infty}(\bar{w})} \leq C_0\|\chi_B\|_{\Lambda^{p_0}(w)}, \quad B \subset X,$$
$$\|T\chi_B\|_{\Lambda^{q_1,\infty}(\bar{w})} \leq C_1\|\chi_B\|_{\Lambda^{p_1}(w)}, \quad B \subset X.$$

Then, if $W, \bar{W} \in \Delta_2$, we have

$$T : \Lambda_X^{p,r}(w) \longrightarrow \Lambda_{\bar{X}}^{q,r}(\bar{w}), \qquad 0 < r \leq \infty,$$

if

$$\frac{1}{p} = \frac{1-\theta}{p_0} + \frac{\theta}{p_1}, \qquad \frac{1}{q} = \frac{1-\theta}{q_0} + \frac{\theta}{q_1}, \qquad 0 < \theta < 1.$$

Proof. If $\bar{W} \in \Delta_2$ there exists $t > 0$ such that $\bar{w} \in B_t$ (see [CGS]). Since the classes B_p are increasing in p, there exists an index $r \in (0,1)$ with $r < p_i$, $i = 0, 1$ and such that $\bar{w} \in B_{q_i/r}$, $i = 0, 1$. Since $\|\chi_B\|_{\Lambda^{p_i}} = C_{p_i,r}\|\chi_B\|_{\Lambda^{p_i,r}}$, $B \subset X$, we have

$$\|T\chi_B\|_{\Lambda^{q_i,\infty}(\bar{w})} \leq C'_i\|\chi_B\|_{\Lambda^{p_i,r}(w)}, \quad B \subset X, \quad i = 0, 1.$$

But $\Lambda^{p_i,r}(w) = \Lambda^r(\tilde{w}_i)$ with $\tilde{w}_i = W^{r/p_i - 1}w$ (Remark 2.2.6) and by Theorem 2.6.9 it follows that

$$T : \Lambda_X^{p_i,r}(w) \longrightarrow \Lambda_{\bar{X}}^{q_i,\infty}(\bar{w}), \qquad i = 0, 1.$$

Applying then Theorem 2.6.5 we obtain the result. □

Remark 2.6.11 (i) Since $\|\chi_B\|_{\Lambda^{p_i}} = C_{p_i,r}\|\chi_B\|_{\Lambda^{p_i,r}}$, $B \subset X$, the previous theorem is still true if we substitute the spaces $\Lambda^{p_i}(w)$ by $\Lambda^{p_i,r_i}(w)$, with $0 < r_i \leq \infty$, $i = 0, 1$.

(ii) An analogous result holds, without the hypothesis $W \in \Delta_2$, if we change the space $\Lambda^{p_1}(w)$ by L^∞, and with $1/p = (1-\theta)/p_0$. This is true since, in this case, one can use Theorem 2.6.2 (where this hypothesis is not needed) to identify the interpolated space.

Chapter 3

The Hardy-Littlewood maximal operator in weighted Lorentz spaces

3.1 Introduction

In the previous chapter, we have introduced and studied the Lorentz spaces $\Lambda_X^{p,q}(w)$. Our purpose in this new chapter is to study the boundedness of the Hardy-Littlewood maximal operator of the type

$$M : \Lambda_u^p(w) \longrightarrow \Lambda_u^p(w), \tag{3.1}$$

and its weak version, $M : \Lambda_u^p(w) \to \Lambda_u^{p,\infty}(w)$ (see below for the definition of these spaces). Our goal is to find necessary and/or sufficient conditions on the weights u and w (in $\mathbb{R}^n$ and $\mathbb{R}^m$ respectively) to have (3.1). We shall also obtain some positive result for the nondiagonal case

$$M : \Lambda_{u_0}^{p_0}(w_0) \longrightarrow \Lambda_{u_1}^{p_1,\infty}(w_1). \tag{3.2}$$

If $w = 1$, (3.1) is equivalent to $M : L^p(u) \longrightarrow L^p(u)$ and this problem was completely solved by Muckenhoupt ([Mu]), who obtained the condition $u \in A_p$, $1 < p < \infty$ (see (3.21)). On the other hand, if $u = 1$, the characterization of (3.1) is equivalent to the boundedness of the Hardy operator $A : L_{\text{dec}}^p(w) \to L^p(w)$ and was obtained ($p > 1$) by Ariño and Muckenhoupt ([AM1]). The condition, in this case, is $w \in B_p$ (see Definition 1.3.1).

The problem (3.2) and the corresponding strong-type boundedness when w_0, w_1 are power weights $(w_i(t) = t^{\alpha_i})$ reduces to $M : L^{p_0,q_0}(u) \to L^{p_1,q_1}(u)$ and was solved, for some cases, by Chung, Hunt, and Kurtz in [CHK, HK] (see also [La]). The complete solution for the general diagonal case (3.1) has been an open question so far, although there are some partial results due to Carro-Soria ([CS3]) and Neugebauer ([Ne2]). In this chapter, we prove (see Theorem 3.3.5) a complete characterization of (3.1) and give necessary and sufficient conditions to have the weak-type boundedness.

The chapter is organized as follows: in section 3.2 we present some general results which will be used in what follows, in section 3.3, we show that (3.1) holds if and only if there exists $q \in (0, p)$ such that

$$\frac{W(u(\bigcup_j Q_j))}{W(u(\bigcup_j E_j))} \leq C \max_j \left(\frac{|Q_j|}{|E_j|}\right)^q,$$

for every finite family of cubes and sets $(Q_j, E_j)_j$, with $E_j \subset Q_j$.

In many cases this condition can be simplified. Among other results, we shall see that although the weight w in (3.1) has to be in some class B_p, the weight u need not be even in A_∞ (see (3.24)). In fact, we shall show that (3.1) can be true with weights u which are not doubling.

In section 3.4, we study the weak boundedness. For example, we prove that if $u_0 = u_1 = u \in A_1$, (3.2) is equivalent to the case $u_0 = u_1 = 1$. Finally, in section 3.5, we completely solve the problem

$$M : L^{p,q}(u) \longrightarrow L^{r,s}(u),$$

and also (3.2), whenever $u_0 = u_1 = u$ is a power weight.

Notation: Letters $u, u_0, u_1, \ldots$, will be used to denote weights in $\mathbb{R}^n$. They will be nonnegative measurable functions, not identically zero and integrable on sets of finite measure. If $A \subset \mathbb{R}^n$ is measurable, we shall denote by $u(A)$ the measure of A in the space $X = (\mathbb{R}^n, u(x)dx)$; that is,

$$u(A) = \int_A u(x)\, dx.$$

The distribution function of $f \in \mathcal{M}(\mathbb{R}^n)$ in this space, will be denoted by λ_f^u and the decreasing rearrangement by f_u^*. $\Lambda_u^{p,q}(w)$ will be the Lorentz space $\Lambda_X^{p,q}(w)$. If $u = 1$ we simply write $\Lambda^{p,q}(w)$. Hence, the "norm" of f in the weighted Lorentz space $\Lambda_u^p(w)$ will be given by

$$\|f\|_{\Lambda_u^p(w)} = \left(\int_0^\infty (f_u^*(t))^p w(t)\, dt\right)^{1/p},$$

and, in the weak space,
$$\|f\|_{\Lambda_u^{p,\infty}(w)} = \sup_{t>0} W^{1/p}(t) f_u^*(t) = \sup_{t>0} t\, W^{1/p}(\lambda_f^u(t)).$$

Let us also recall that w is a weight in $\mathbb{R}^+$ with support in $[0, u(\mathbb{R}^n)]$ and that
$$0 \leq W(t) = \int_0^t w(s)\, ds < \infty,\ t > 0.$$
Finally, the Hardy-Littlewood maximal operator M is defined by
$$Mf(x) = \sup_{x \in Q} \frac{1}{|Q|} \int_Q |f(y)|\, dy, \qquad x \in \mathbb{R}^n,\ f \in \mathcal{M}(\mathbb{R}^n),$$
where the supremum extends over all cubes Q, with sides parallel to the axes.

3.2 Some general results

It is known that the decreasing rearrangement of Mf, with respect to the Lebesgue measure, is equivalent ([BS]) to the function f^{**}:
$$(Mf)^*(t) \approx f^{**}(t) = Af^*(t), \qquad t > 0. \tag{3.3}$$

Since every decreasing and positive function in $\mathbb{R}^+$ is equal a.e. to the decreasing rearrangement of a measurable function in $\mathbb{R}^n$, we deduce that the boundedness of $M : \Lambda^p(w) \to \Lambda^{p,\infty}(w)$ (resp. $M : \Lambda^p(w) \to \Lambda^p(w)$) is equivalent to the boundedness of $A : L_{\text{dec}}^p(w) \to L^{p,\infty}(w)$ (resp. $A : L_{\text{dec}}^p(w) \to L^p(w)$). Therefore (see section 1.3) a necessary and sufficient condition is $w \in B_{p,\infty}$ (resp. $w \in B_p$). On the other hand, when $w = 1$, the boundedness $M : \Lambda_u^p(w) \to \Lambda_u^{p,\infty}(w)$ is equivalent to $M : L^p(u) \to L^{p,\infty}(u)$ which is known to be $u \in A_p$ (for $p \geq 1$), the Muckenhoupt class of weights ([Mu, MW, CF]). This motivates the following definition:

Definition 3.2.1 If $0 < p < \infty$, we write $w \in B_p(u)$ (respectively $w \in B_{p,\infty}(u)$) if the boundedness $M : \Lambda_u^p(w) \to \Lambda_u^p(w)$ (resp. $M : \Lambda_u^p(w) \to \Lambda_u^{p,\infty}(w)$) holds. We shall also write $u \in A_p(w)$ if $M : \Lambda_u^p(w) \to \Lambda_u^{p,\infty}(w)$ holds.

That is, $u \in A_p(w) \Leftrightarrow w \in B_{p,\infty}(u)$. It is clear then that $A_p(1) = A_p$, $1 \leq p < \infty$, and that $B_p(1) = B_p$, $B_{p,\infty}(1) = B_{p,\infty}$, $0 < p < \infty$. On the other hand, since $\Lambda^p \subset \Lambda^{p,\infty}$, we always have that $B_p(u) \subset B_{p,\infty}(u)$, $0 < p < \infty$. With this terminology, our purpose will be to identify the classes $B_p(u)$ and $B_{p,\infty}(u)$, or equivalently to identify the classes $A_p(w)$.

Using (3.3), the nondiagonal case (with $u = 1$) is also very simple: the boundedness of $M : \Lambda^{p_0,q_0}(w_0) \to \Lambda^{p_1,q_1}(w_1)$ is equivalent to the boundedness of the Hardy operator $A : L_{\text{dec}}^{p_0,q_0}(w_0) \to L^{p_1,q_1}(w_1)$, which is known in most of the cases.

The natural question now is to see if there exists an analogue to (3.3) when the rearrangement f^* is taken with respect to a weight $u \neq 1$; that is, if $(Mf)_u^*(t) \approx Af_u^*(t)$ or $(Mf)_u^*(t) \leq CAf_u^*(t)$, when u is a weight in $\mathbb{R}^n$ satisfying certain conditions. However, it was shown in [CS3] (see also [LN1, LN2]) that $(Mf)_u^* \approx Af_u^*$ if and only if $u \approx 1$, and $(Mf)_u^* \leq C(Af_u^{*p})^{1/p}$ if and only if $M : L^p(u) \to L^{p,\infty}(u)$, $p > 1$ (equivalently if $u \in A_p$). The following theorem generalizes this result, characterizing $M : \Lambda_u^p(w) \to \Lambda_u^{p,\infty}(w)$ in terms of an expression like (3.3).

Theorem 3.2.2 *If $0 < p < \infty$, $M : \Lambda_u^p(w) \to \Lambda_u^{p,\infty}(w)$ is bounded if and only if*

$$(Mf)_u^*(t) \leq C\left(\frac{1}{W(t)} \int_0^t (f_u^*)^p(s) w(s) \, ds\right)^{1/p}, \qquad t > 0, \; f \in \mathcal{M}(\mathbb{R}^n).$$

Proof. The inequality of the statement implies

$$W^{1/p}(t)(Mf)_u^*(t) \leq C\left(\int_0^t (f_u^*)^p(s) w(s) \, ds\right)^{1/p} \leq C\|f\|_{\Lambda_u^p(w)}, \qquad t > 0,$$

or, equivalently, $\|Mf\|_{\Lambda_u^{p,\infty}(w)} \leq C\|f\|_{\Lambda_u^p(w)}$. This proves that this condition is sufficient. On the other hand, let us assume that $M : \Lambda_u^p(w) \to \Lambda_u^{p,\infty}(w)$ is bounded and let $f \in \Lambda_u^p(w)$. If $f = f_0 + f_1$, with $f_1 \in L^\infty$, we have, for every $t > 0$,

$$\begin{aligned}(Mf)_u^*(t) &\leq (Mf_0)_u^*(t) + (Mf_1)_u^*(0) \\ &\leq W^{-1/p}(t)\|Mf_0\|_{\Lambda_u^{p,\infty}(w)} + \|Mf_1\|_{L^\infty(u)} \\ &\leq CW^{-1/p}(t)(\|f_0\|_{\Lambda_u^p(w)} + W^{1/p}(t)\|f_1\|_{L^\infty(u)}),\end{aligned}$$

and taking the infimum over all decompositions $f = f_1 + f_0$ we obtain

$$(Mf)_u^*(t) \leq CW^{-1/p}(t) K(f, W^{1/p}(t), \Lambda_u^p(w), L^\infty(u)), \qquad t > 0. \qquad (3.4)$$

By Theorem 2.6.2, $K(f_u^*, W^{1/p}(t), L^p(w), L^\infty(w))$ is equivalent to the K-functional and therefore (see [BL]) equivalent to

$$\left(\int_0^{W(t)} ((f_u^*)_w^*)^p(s)\,ds\right)^{1/p} = \left(\int_0^t (f_u^*)^p(s) w(s)\,ds\right)^{1/p}.$$

This and (3.4) give us the inequality we are looking for. $\square$

Remark 3.2.3 The same argument works for the nondiagonal case and hence (3.2) is equivalent, if $0 < p_0, p_1 < \infty$, to the inequality

$$(Mf)_{u_1}^*(t) \leq C\left(\frac{1}{W_1^{p_0/p_1}(t)} \int_0^{W_0(t)} ((f_{u_0}^*)_{w_0}^*)^{p_0}(s)\,ds\right)^{1/p_0}, \quad t > 0.$$

Next result gives a necessary and sufficient condition to have the weak-type inequality,

$$\|Mf\|_{\Lambda_{u_1}^{p_1,\infty}(w_1)} \leq C\|f\|_{\Lambda_{u_0}^{p_0}(w_0)}$$

on characteristic functions $f = \chi_E$, $E \subset \mathbb{R}^n$.

Theorem 3.2.4 Let $0 < p_0, p_1 < \infty$. Then,

$$\|M\chi_E\|_{\Lambda_{u_1}^{p_1,\infty}(w_1)} \leq C\|\chi_E\|_{\Lambda_{u_0}^{p_0}(w_0)}, \qquad E \subset \mathbb{R}^n \text{ measurable}, \tag{3.5}$$

if and only if, for every finite family of cubes $(Q_j)_{j=1}^J$ and every family of measurable sets $(E_j)_{j=1}^J$, with $E_j \subset Q_j$, for every j, we have that

$$\frac{W_1^{1/p_1}(u_1(\bigcup_j Q_j))}{W_0^{1/p_0}(u_0(\bigcup_j E_j))} \leq C \max_j \frac{|Q_j|}{|E_j|}. \tag{3.6}$$

Proof. To prove the necessary condition, let us consider $f = \chi_E$ with $E = \bigcup_j E_j$. If $t > 0$ is such that $1/t > \max_j \frac{|Q_j|}{|E_j|}$, then

$$\frac{1}{|Q_j|} \int_{Q_j} f(x)\,dx \geq \frac{|E_j|}{|Q_j|} > t,$$

and we have that $Q_j \subset \{Mf > t\}$. This holds for every $j = 1, \ldots, J$ and hence, $\bigcup_j Q_j \subset \{Mf > t\}$. Therefore,

$$tW_1^{1/p_1}(u_1(\bigcup_j Q_j)) \leq tW_1^{1/p_1}(\lambda_{Mf}^{u_1}(t)) \leq C\|f\|_{\Lambda_{u_0}^{p_0}(w_0)} = CW_0^{1/p_0}(u_0(\bigcup_j E_j)),$$

and consequently
$$\frac{W_1^{1/p_1}(u_1(\bigcup_j Q_j))}{W_0^{1/p_0}(u_0(\bigcup_j E_j))} \leq \frac{C}{t}.$$

Since $1/t > \max_j \frac{|Q_j|}{|E_j|}$ is arbitrary, we have shown (3.6).

Conversely, let us assume that (3.6) holds. Then, this inequality also holds if the families $(Q_j)_j$, $(E_j)_j$ are countable nonfinite. If $f = \chi_E \in \mathcal{M}(\mathbb{R}^n)$ and $t > 0$, for every $x \in \{Mf > t\}$, there exists a cube Q, with $x \in Q$ and such that $\int_Q f(x)\,dx = |E \cap Q| > t$. By definition we have that $Mf(y) > t$, for every $y \in Q$ and hence, $Q \subset \{Mf > t\}$. Using again this argument, we obtain a countable family of cubes $(Q_j)_j$ such that $\{Mf > t\} = \bigcup_j Q_j$ and satisfying (with $E_j = E \cap Q_j$) $|E_j|/|Q_j| > t$, $j = 1, 2, \ldots$ Then, $t \leq \inf_j \frac{|E_j|}{|Q_j|} = (\sup_j \frac{|Q_j|}{|E_j|})^{-1}$ and we have, applying (3.6),

$$\begin{aligned}tW_1^{1/p_1}(\lambda_{Mf}^{u_1}(t)) &\leq \frac{W_1^{1/p_1}(u_1(\bigcup_j Q_j))}{\sup_j \frac{|Q_j|}{|E_j|}} \\ &\leq CW_0^{1/p_0}(u_0(\bigcup_j E_j)) \leq CW_0^{1/p_0}(u_0(E)).\end{aligned}$$

Taking now the supremum in t we obtain that

$$\|Mf\|_{\Lambda_{u_1}^{p_1,\infty}(w_1)} \leq CW_0^{1/p_0}(u_0(E)) = C\|f\|_{\Lambda_{u_0}^{p_0}(w_0)}. \qquad \square$$

Remark 3.2.5 In the condition (3.6), we can assume that the sets $(E_j)_j$ are disjoint. In fact, for every finite family of cubes $(Q_j)_j$ there exists a subfamily $(Q_{j_k})_k$ of disjoint cubes such that $\bigcup_j Q_j \subset \bigcup_k Q_{j_k}^*$, where every $Q_{j_k}^*$ is a dilation of Q_{j_k} with side three times bigger (see for example [Stn]). The sets $(E_{j_k})_k$ are then disjoint and, if the condition holds, we have

$$\frac{W_1^{1/p_1}(u_1(\bigcup_j Q_j))}{W_0^{1/p_0}(u_0(\bigcup_j E_j))} \leq \frac{W_1^{1/p_1}(u_1(\bigcup_k Q_{j_k}^*))}{W_0^{1/p_0}(u_0(\bigcup_k E_{j_k}))} \leq C \max_k \frac{|Q_{j_k}^*|}{|E_{j_k}|}$$
$$\leq CC_n \max_k \frac{|Q_{j_k}|}{|E_{j_k}|} \leq CC_n \max_j \frac{|Q_j|}{|E_j|}.$$

Corollary 3.2.6 If $M : \Lambda_{u_0}^{p_0}(w_0) \to \Lambda_{u_1}^{p_1,\infty}(w_1)$, $0 < p_0, p_1 < \infty$, then

$$\frac{W_1^{1/p_1}(u_1(Q))}{|Q|} \leq C \frac{W_0^{1/p_0}(u_0(E))}{|E|}, \quad E \subset Q,$$

for every cube $Q \subset \mathbb{R}^n$. In particular $W_0(t) > 0$, $t > 0$, and $u_0(x) > 0$ a.e. $x \in \mathbb{R}^n$.

The following two propositions are a consequence of these results.

Proposition 3.2.7 Let $0 < p_0, p_1 < \infty$ and let us assume that $M : \Lambda_{u_0}^{p_0}(w_0) \to \Lambda_{u_1}^{p_1,\infty}(w_1)$, then $u_0 \notin L^1(\mathbb{R}^n)$.

Proof. By Corollary 3.2.6 we have that, for every cube Q,

$$\frac{|E|}{|Q|} \leq C \frac{W_0^{1/p_0}(u_0(E))}{W_1^{1/p_1}(u_1(Q))}, \quad E \subset Q. \tag{3.7}$$

Let us assume that $u_0(\mathbb{R}^n) < \infty$. Then, if $a \in (0, u_1(\mathbb{R}^n))$ and $b \in (0,1)$ are such that

$$C \frac{W_0^{1/p_0}(bu_0(\mathbb{R}^n))}{W_1^{1/p_1}(a)} < 5^{-n},$$

with C as in (3.7), we have that, if $u_1(Q) \geq a$, the inequality $u_0(E)/u_0(Q) \leq b$ implies

$$\frac{|E|}{|Q|} \leq C \frac{W_0^{1/p_0}(u_0(E))}{W_1^{1/p_1}(u_1(Q))} \leq C \frac{W_0^{1/p_0}(bu_0(\mathbb{R}^n))}{W_1^{1/p_1}(a)} < 5^{-n};$$

that is, if Q is an arbitrary cube with $u_1(Q) \geq a$,

$$E \subset Q, \ |E| \geq 5^{-n}|Q| \Rightarrow u_0(E) > bu_0(Q). \tag{3.8}$$

Let now Q_0 be a cube with $u_1(3Q_0) \geq a$ (for each cube Q, kQ denotes another cube with the same center Q and side k times bigger than Q). Let $\tilde{Q} \subset 3Q_0$ be a cube whose interior is disjoint with Q_0 and such that $|\tilde{Q}| = |Q_0|$. Then, $5\tilde{Q} \supset 3Q_0$ and hence, $u_1(5\tilde{Q}) \geq a$, and we have by (3.8)

$$u_0(\tilde{Q}) > bu_0(5\tilde{Q}) \geq bu_0(Q_0).$$

Therefore, $u_0(3Q_0) \geq u_0(Q_0) + u_0(\tilde{Q}) \geq (1+b)u_0(Q_0) = \alpha u_0(Q_0)$, where $\alpha = 1+b > 1$. By the same argument, $u_0(9Q_0) \geq \alpha u_0(3Q_0) \geq \alpha^2 u_0(Q_0)$ and, in general, $u_0(3^n Q_0) \geq \alpha^n u_0(Q_0)$. Since $\lim_n u_0(3^n Q_0) = u_0(\mathbb{R}^n)$ and $\alpha > 1$ we have that $u_0(\mathbb{R}^n) = \infty$ (observe that $u_0(Q_0) > 0$ by (3.7)) contradicting the initial assumption. $\square$

Proposition 3.2.8 Let $0 < p_0, p_1 < \infty$. If $M : \Lambda_u^{p_0}(w) \to \Lambda_u^{p_1,\infty}(w)$, then $p_1 \leq p_0$. If, in addition, $w \notin L^1(\mathbb{R}^+)$, then $p_1 = p_0$.

Proof. By Corollary 3.2.6, we have that
$$W^{1/p_1 - 1/p_0}(u(Q)) \leq C,$$
for every cube $Q \subset \mathbb{R}^n$, and by Proposition 3.2.7
$$W^{1/p_1 - 1/p_0}(r) \leq C, \quad r > 0.$$
Since $\lim_{t \to 0} W(t) = 0$, we obtain that $p_1 \leq p_0$. If $w \notin L^1(\mathbb{R}^+)$, $\lim_{t \to \infty} W(t) = \infty$ and the previous inequality holds only if $p_0 = p_1$. □

Proposition 3.2.9 Let $0 < p, q, r < \infty$. If $M : \Lambda_u^{p,q}(w) \to \Lambda_u^{p,r}(w)$, then $r \geq q$.

Proof. $|f| \leq Mf$ for every $f \in \mathcal{M}(\mathbb{R}^n)$ and the hypothesis implies that $\Lambda_u^{p,q}(w) \subset \Lambda_u^{p,r}(w)$. Since $(\mathbb{R}^n, u(x)dx)$ and $(\mathbb{R}^+, w(t)dt)$ are nonatomic spaces, and $\|f\|_{\Lambda_u^{p,q}(w)} = \|f_u^*\|_{L^{p,q}(w)}$, the previous embedding implies that
$$\left(\int_0^b g^r(t) t^{r/p - 1} dt \right)^{1/r} \leq C \left(\int_0^b g^q(t) t^{q/p - 1} dt \right)^{1/q}, \quad g \downarrow,$$
with $b = W(u(\mathbb{R}^n))$. Equivalently,
$$\sup_{g \downarrow} \frac{\int_0^b g(t) t^{r/p - 1} dt}{\left(\int_0^b g(t)^{q/r} t^{q/p - 1} dt \right)^{r/q}} < \infty.$$
Now, by Theorem 1.3.5, if this supremum is finite, then $r \geq q$. □

The following result is a consequence of the interpolation theorems developed in section 1.2 of the previous chapter.

Theorem 3.2.10 Let $0 < p_0, p_1 < \infty$ and let us assume that $W_1 \in \Delta_2$. Then, the boundedness $M : \Lambda_{u_0}^{p_0}(w_0) \to \Lambda_{u_1}^{p_1,\infty}(w_1)$ on characteristic functions:
$$\|M\chi_E\|_{\Lambda_{u_1}^{p_1,\infty}(w_1)} \leq C \|\chi_E\|_{\Lambda_{u_0}^{p_0}(w_0)}, \quad E \subset \mathbb{R}^n,$$
implies
$$M : \Lambda_{u_0}^{q_0, r}(w_0) \longrightarrow \Lambda_{u_1}^{q_1, r}(w_1), \quad 0 < r \leq \infty,$$
for $p_i < q_i < \infty$, $i = 0, 1$, $p_1/p_0 = q_1/q_0$. In particular, we have that $M : \Lambda_{u_0}^{q_0}(w_0) \to \Lambda_{u_1}^{q_1}(w_1)$ if $p_1 \geq p_0$.

Proof. Since $M : L^\infty \to L^\infty$, the statement is an immediate consequence of Theorem 2.6.10 (see also Remark 2.6.11). □

Combining Theorem 3.2.4 with the results of section 2.6 of the previous chapter, on boundedness of continuous order operators on Lorentz spaces, we obtain the following characterization for the weak-type boundedness in the case $0 < p_0 \le 1$ and $w_1 \in B_{p_1/p_0}$.

Theorem 3.2.11 *If $0 < p_0 \le 1$, $0 < p_1 < \infty$, and $w_1 \in B_{p_1/p_0}$ we have the boundedness $M : \Lambda^{p_0}_{u_0}(w_0) \to \Lambda^{p_1,\infty}_{u_1}(w_1)$ if and only if, for every finite family of cubes $(Q_j)_{j=1}^J$ and every family of measurable sets $(E_j)_{j=1}^J$ with $E_j \subset Q_j$, for every j, we have that*

$$\frac{W_1^{1/p_1}(u_1(\bigcup_j Q_j))}{W_0^{1/p_0}(u_0(\bigcup_j E_j))} \le C \max_j \frac{|Q_j|}{|E_j|}.$$

Proof. Since M is an order continuous operator (Definition 1.2.3) on the regular class $L = \Lambda^{p_0}_{u_0}(w_0)$, we have, by Theorem 2.6.9, that the boundedness is equivalent to (3.5) and thus, condition (3.6) is necessary and sufficient. □

3.3 Strong-type boundedness in the diagonal case

In this section we shall give a characterization of the $B_p(u)$ class in the more general case; that is, we shall obtain a necessary and sufficient condition to have the boundedness

$$M : \Lambda^p_u(w) \longrightarrow \Lambda^p_u(w).$$

To this end, we first need the two following technical lemmae:

Lemma 3.3.1 *Let $0 < p < \infty$ and let us assume that, for every cube $Q \subset \mathbb{R}^n$,*

$$\frac{W(u(Q))}{|Q|^p} \le C \frac{W(u(E))}{|E|^p}, \qquad E \subset Q,$$

with C independent of Q. Then, $w \in B_q$ for every $q > p$. In particular $W \in \Delta_2$.

Proof. The measure $u(x)dx$ is σ-finite and nonatomic and, therefore, $(\mathbb{R}^n, u(x)dx)$ is a resonant measure space. In these spaces, it holds that

$$\int_0^\infty f^*(s)g^*(s)\,ds = \sup_{h^*=g^*} \int f(x)h(x)u(x)\,dx$$

(see [BS]) for every measurable functions f, g. In our case, we have that, for $0 < t < u(Q)$,

$$\begin{aligned}(u^{-1}\chi_Q)_u^{**}(t) &= \frac{1}{t}\int_0^t (u^{-1}\chi_Q)_u^*(s)\,ds \\ &= \frac{1}{t}\sup\left\{\int_Q u^{-1}(x)\chi_E(x)\,u(x)\,dx : u(E)=t,\ E \subset Q\right\} \\ &= \frac{1}{t}\sup\{|E| : u(E)=t,\ E \subset Q\} \\ &\leq C\frac{|Q|}{W^{1/p}(u(Q))}\frac{W^{1/p}(t)}{t}.\end{aligned}$$

Since the function $(u^{-1}\chi_Q)_u^{**}$ is decreasing and,

$$(u^{-1}\chi_Q)_u^{**}(t) \geq (u^{-1}\chi_Q)_u^{**}(u(Q)) = \frac{|Q|}{u(Q)},$$

we obtain that

$$\frac{W^{1/p}(u(Q))}{u(Q)} \leq C\frac{W^{1/p}(t)}{t}, \qquad 0 < t < u(Q),$$

which is equivalent to $\frac{W^{1/p}(r)}{r} \leq C\frac{W^{1/p}(t)}{t}$, $0 < t < r < \infty$ and hence (see, for example [So]) $w \in \bigcup_{q>p} B_q$. $\square$

We shall also need the following result due to Hunt-Kurtz ([HK]).

Lemma 3.3.2 *If $t > 0$, $E \subset \mathbb{R}^n$, and we denote by $E_t = \{M\chi_E > t\}$, there exists a constant $\alpha > 1$, depending only on the dimension, such that*

$$(E_t)_s \supset E_{\alpha ts}, \qquad s, t \in (0, 1).$$

It is known that the boundedness of $M : L^p(u) \to L^p(u)$, $p > 1$, implies $M : L^{p-\epsilon}(u) \to L^{p-\epsilon}(u)$ for some $\epsilon > 0$. Analogously, $M : \Lambda^p(w) \to \Lambda^p(w)$ is equivalent to $w \in B_p$, that implies $w \in B_{p-\epsilon}$. As we are going to see next, this also holds for the general case $M : \Lambda^p_u(w) \to \Lambda^p_u(w)$; that is, if $w \in B_p(u)$, $0 < p < \infty$, there exists $\epsilon > 0$ such that $w \in B_{p-\epsilon}(u)$. In fact, in our next theorem, we prove a stronger result: if $M : \Lambda^p_u(w) \to \Lambda^p_u(w)$ on characteristics functions, then $w \in B_{p-\epsilon}$. Part of its proof follows the same patterns developed in Theorem 2 of [HK].

Theorem 3.3.3 *Let $0 < p, r < \infty$ and let us assume that*

$$\|M\chi_E\|_{\Lambda^{p,r}_u(w)} \leq C\|\chi_E\|_{\Lambda^p_u(w)}, \qquad E \subset \mathbb{R}^n.$$

Then, there exists $q \in (0, p)$ such that $M : \Lambda^q_u(w) \to \Lambda^q_u(w)$.

Proof. First, we shall prove that the inequality of the statement is also true if we substitute the indices p, r by some others $\tilde{p}$ and $\tilde{r}$ with $\tilde{p} < p$ and $\tilde{r} < r$. To this end, let us observe that the hypothesis of the theorem is equivalent (c.f. Proposition 2.2.5) to the inequality

$$\int_0^1 t^{r-1} W^{r/p}(u(E_t))\, dt \leq B W^{r/p}(u(E)), \qquad E \subset \mathbb{R}^n, \qquad (3.9)$$

with E_t, $t > 0$, defined as in the previous lemma, and B is a constant independent of E. Let $\alpha > 1$, the constant, depending only on the dimension, of Lemma 3.3.2. We shall prove that, for every $n = 0, 1, 2, \ldots$, the inequality

$$\int_0^1 t^{r-1} W^{r/p}(u(E_t)) \frac{1}{n!} \log^n \frac{1}{t}\, dt \leq B(B\alpha^r)^n W^{r/p}(u(E)), \qquad E \subset \mathbb{R}^n, \qquad (3.10)$$

holds. If $n = 0$, (3.10) is (3.9). By induction, we only have to show that (3.10) implies an analogous inequality with $n+1$ instead of n. To see this, and since $s \in (0, 1)$, we apply (3.10) to the set $E = E_s$, obtaining

$$\int_0^1 t^{r-1} W^{r/p}(u((E_s)_t)) \frac{1}{n!} \log^n \frac{1}{t}\, dt \leq B(B\alpha^r)^n W^{r/p}(u(E_s)). \qquad (3.11)$$

By Lemma 3.3.2, the left hand side of (3.11) is greater than or equal to

$$\int_0^{1/\alpha} t^{r-1} W^{r/p}(u(E_{\alpha s t})) \frac{1}{n!} \log^n \frac{1}{t}\, dt$$

$$= (\alpha s)^{-r} \int_0^s x^{r-1} W^{r/p}(u(E_x)) \frac{1}{n!} \log^n \frac{\alpha s}{x}\, dx$$

$$\geq (\alpha s)^{-r} \int_0^s x^{r-1} W^{r/p}(u(E_x)) \frac{1}{n!} \log^n \frac{s}{x}\, dx.$$

Thus, from (3.11) it follows that

$$\frac{1}{s}\int_0^s x^{r-1}W^{r/p}(u(E_x))\frac{1}{n!}\log^n\frac{s}{x}\,dx \leq (B\alpha^r)^{n+1}s^{r-1}W^{r/p}(u(E_s)).$$

Integrating in $s \in (0,1)$ both members of this inequality and by (3.9), we deduce that

$$\int_0^1 \frac{1}{s}\int_0^s x^{r-1}W^{r/p}(u(E_x))\frac{1}{n!}\log^n\frac{s}{x}\,dx\,ds \leq B(B\alpha^r)^{n+1}W^{r/p}(u(E)),$$

and changing the order of integration in the left hand side, we obtain

$$\int_0^1 x^{r-1}W^{r/p}(u(E_x))\frac{1}{(n+1)!}\log^{n+1}\frac{1}{x}\,dx \leq B(B\alpha^r)^{n+1}W^{r/p}(u(E)),$$

as we wanted to prove.

Let now $R \in (0,1)$. The inequality (3.10) can be written in the following form:

$$\int_0^1 t^{r-1}W^{r/p}(u(E_t))\frac{(\frac{R}{B\alpha^r}\log\frac{1}{t})^n}{n!}\,dt \leq BR^n W^{r/p}(u(E)),$$

and summing in n we obtain, with $\delta = R/(B\alpha^r) > 0$,

$$\int_0^1 t^{r-\delta-1}W^{r/p}(u(E_t))\,dt \leq \frac{B}{1-R}W^{r/p}(u(E)),$$

equivalently

$$\|M\chi_E\|_{\Lambda_u^{\tilde{p},\tilde{s}}(w)} \leq \tilde{C}\|\chi_E\|_{\Lambda_u^{\tilde{p}}(w)}, \qquad E \subset \mathbb{R}^n,$$

where $\tilde{p} = p(r-\delta)/r < p$, $\tilde{s} = r - \delta$ (see Proposition 2.2.5). In particular,

$$\|M\chi_E\|_{\Lambda_u^{\tilde{p},\infty}(w)} \leq \tilde{C}\|\chi_E\|_{\Lambda_u^{\tilde{p}}(w)}, \qquad E \subset \mathbb{R}^n.$$

Also, by Theorem 3.2.4 and Lemma 3.3.1 we have that $W \in \Delta_2$. Then, we can apply Theorem 3.2.10 (with $u_0 = u_1 = u$, $w_0 = w_1 = w$, $p_0 = p_1 = \tilde{p}$) to conclude the result. □

An immediate consequence of Theorem 3.3.3 and Lemma 3.3.1 is the following:

Corollary 3.3.4 Let $0 < p < \infty$ and let u be an arbitrary weight in $\mathbb{R}^n$. Then,

1. If $w \in B_p(u)$ there exists $q < p$ such that $w \in B_q(u)$.

2. $B_p(u) \subset B_p$, that is, the boundedness $M : \Lambda_u^p(w) \to \Lambda_u^p(w)$ implies $M : \Lambda^p(w) \to \Lambda^p(w)$.

We can now obtain the characterization of the boundedness $M : \Lambda_u^p(w) \to \Lambda_u^p(w)$ or equivalently of the classes $B_p(u)$ for every u.

Theorem 3.3.5 *Let u, w be weights in $\mathbb{R}^n$ and $\mathbb{R}^+$ respectively. If $0 < p < \infty$ the following results are equivalent:*

(i) $M : \Lambda_u^p(w) \to \Lambda_u^p(w)$.

(ii) $\|M\chi_E\|_{\Lambda_u^p(w)} \leq C\|\chi_E\|_{\Lambda_u^p(w)}$, $\quad E \subset \mathbb{R}^n$.

(iii) $M : \Lambda_u^q(w) \to \Lambda_u^q(w)$, with $q \in (0, p)$.

(iv) *There exists $q \in (0, p)$ such that* $\|M\chi_E\|_{\Lambda_u^{q,\infty}(w)} \leq C\|\chi_E\|_{\Lambda_u^q(w)}$, $E \subset \mathbb{R}^n$.

(v) *There exists $q \in (0, p)$ such that, for every finite family of cubes $(Q_j)_{j=1}^J$ and every family of measurable sets $(E_j)_{j=1}^J$ with $E_j \subset Q_j$, for every j, we have that*
$$\frac{W(u(\bigcup_j Q_j))}{W(u(\bigcup_j E_j))} \leq C \max_j \left(\frac{|Q_j|}{|E_j|}\right)^q. \qquad (3.12)$$

(vi) $((M\chi_E)_u^*(t))^q \leq C \dfrac{W(u(E))}{W(t)}$, $t > 0$, $E \subset \mathbb{R}^n$, with $q \in (0, p)$ independent of t and E.

Proof.
(i) $\Rightarrow$ (ii) is immediate and (ii)$\Rightarrow$(iii) is given in Theorem 3.3.3. (iii)$\Rightarrow$(iv) is also immediate and (iv)$\Rightarrow$(i) is a consequence of Theorem 3.2.10 (since $W \in \Delta_2$ by Theorem 3.2.4 and Lemma 3.3.1). The equivalence (iv)$\Leftrightarrow$(v) is Theorem 3.2.4. On the other hand, (vi) implies, for $E \subset \mathbb{R}^n$,

$$\|M\chi_E\|_{\Lambda_u^{q,\infty}(w)} = \sup_{t>0} W^{1/q}(t)(M\chi_E)_u^*(t) \leq CW^{1/q}(u(E)) = C\|\chi_E\|_{\Lambda_u^q(w)},$$

which is condition (iv). Finally, (iii) implies $M : \Lambda_u^q(w) \to \Lambda_u^{q,\infty}(w)$ and by Theorem 3.2.2, we get (vi). $\square$

Remark 3.3.6

As we mentioned in Remark 3.2.5, we can assume that the sets $(E_j)_j$ in (3.12) are disjoint. If the weight u is doubling (i.e., $u(2Q) \leq Cu(Q)$), then also the cubes $(Q_j)_j$ can be taken disjoint.

Let us assume that the weight w satisfies the following property: for every $\alpha > 1$ there exists a constant C_α such that

$$\frac{W(\sum_j r_j)}{W(\sum_j t_j)} \leq C_\alpha \max_j \left(\frac{W(r_j)}{W(t_j)}\right)^\alpha, \qquad (3.13)$$

for every finite family of positive numbers $\{(r_j, t_j)\}_{j=1}^m$, with $0 < t_j < r_j$, $j = 1, \ldots, m$. Then, we only need to check condition (3.12) for a unique cube Q_j and a unique set E_j, that is, it is equivalent to the inequality

$$\frac{W(u(Q))}{|Q|^q} \leq C \frac{W(u(E))}{|E|^q}, \qquad E \subset Q, \qquad (3.14)$$

for $q < p$, and for every cube $Q \subset \mathbb{R}^n$. To see this, observe that this condition is a consequence of (3.12) and, if (3.14) holds and $(E_j)_j$ is a disjoint family with $E_j \subset Q_j$, then

$$\begin{aligned}\frac{W(u(\bigcup_j Q_j))}{W(u(\bigcup_j E_j))} &\leq \frac{W(\sum_j u(Q_j))}{W(\sum_j u(E_j))} \leq C_\alpha \max_j \left(\frac{W(u(Q_j))}{W(u(E_j))}\right)^\alpha \\ &\leq CC_\alpha \max_j \left(\frac{|Q_j|}{|E_j|}\right)^{\alpha q},\end{aligned}$$

and it is enough to take $\alpha > 1$ with $\alpha q < p$. We observe that every power weight $w(t) = t^\alpha$, $\alpha > -1$, satisfies the above condition.

If $w = 1$, $p > 1$, condition (3.12) says that $u \in A_p$, since this last condition is equivalent (see, for example [Stn]) to the existence of $q \in (1, p)$ such that

$$\frac{u(Q)}{|Q|^q} \leq C \frac{u(E)}{|E|^q}, \qquad E \subset Q,$$

for every cube Q, which is equivalent to (3.12) (by (ii)).

If $u = 1$, then it is immediate to see that (3.12) is equivalent to $w \in B_p$.

This last observation can be generalized in the following way.

Theorem 3.3.7 *If $u \in A_1$ then,*

$$M : \Lambda_u^p(w) \to \Lambda_u^p(w) \Leftrightarrow M : \Lambda^p(w) \to \Lambda^p(w), \qquad 0 < p < \infty.$$

With more generality, if $0 < p < \infty$, then $B_p(u) = B_p$ if and only if $u \in \bigcap_{q>1} A_q$.

Proof. We already know, by Corollary 3.3.4, that $B_p(u) \subset B_p$. On the other hand, if $w \in B_p$, there exists $l < p$ such that $w \in B_l$ and we have that $W(r)/W(t) \leq C(r/t)^l$, $0 < t < r < \infty$. Let $s > 1$ be such that $sl < p$. If $u \in \bigcap_{q>1} A_q$, in particular $u \in A_s$, and it holds that $u(Q)/u(E) \leq C(|Q|/|E|)^s$, $E \subset Q$. Therefore, for every family $(Q_j)_j$ of cubes and $E_j \subset Q_j$ (pairwise disjoint),

$$\frac{W(u(\bigcup_j Q_j))}{W(u(\bigcup_j E_j))} \leq \frac{W(\sum_j u(Q_j))}{W(\sum_j u(E_j))} \leq C\Big(\frac{\sum_j u(Q_j)}{\sum_j u(E_j)}\Big)^l$$
$$\leq C \max_j \Big(\frac{u(Q_j)}{u(E_j)}\Big)^l \leq C \max_j \Big(\frac{|Q_j|}{|E_j|}\Big)^{sl},$$

which implies that $w \in B_p(u)$ (Theorem 3.3.5 (v)).

Conversely, let us assume now that $B_p(u) = B_p$. For every $q < p$ the weight $w(t) = t^{q-1}$ is in $B_p = B_p(u)$ and by (3.12) we have that, with $q_1 \in (q, p)$,

$$\Big(\frac{u(Q)}{u(E)}\Big)^q = \frac{W(u(Q))}{W(u(E))} \lesssim \Big(\frac{|Q|}{|E|}\Big)^{q_1},$$

and thus,

$$\frac{u(Q)}{|Q|^{q_1/q}} \leq C \frac{u(E)}{|E|^{q_1/q}}, \qquad E \subset Q,$$

for every cube Q and hence, if $r = q_1/q \in (1, p/q)$, we obtain $u \in \bigcap_{s>r} A_s$. Since this argument works for $q < p$, we deduce that $u \in \bigcap_{s>1} A_s$. $\square$

Remark 3.3.8 (i) We know that the boundedness of $M : \Lambda^p(w) \to \Lambda^p(w)$ holds if and only if $w \in B_p$ and, on the other hand, $M : \Lambda_u^p \to \Lambda_u^p$, $p > 1$ (that is, $M : L^p(u) \to L^p(u)$) if and only if $u \in A_p$. Then, one could think that the boundedness $M : \Lambda_u^p(w) \to \Lambda_u^p(w)$, $p > 1$, holds if and only if $u \in A_p$, $w \in B_p$, that is, $B_p(u) = B_p$ if $u \in A_p$. The previous result shows that this is not true. In fact, as a consequence of it, $B_p(u) \neq B_p$, $1 < p < \infty$, if $u \in A_p \setminus A_q$ with $1 < q < p$.

(ii) The fact that $B_p \subset B_p(u)$ if $u \in \bigcap_{q>1} A_q$ was already proved in [CS3] and [Ne2], for the case $p \geq 1$.

Using the same idea one can analogously prove the following result which together with Theorem 3.3.7 improves Corollary 3.3 in [CS3] and Theorem 4.1 in [Ne2], to consider the whole range $0 < q < \infty$.

Theorem 3.3.9 *If $1 < p < \infty$ and $u \in A_p$, then $B_{q/p,\infty} \subset B_q(u)$, $0 < q < \infty$.*

By Proposition 3.2.7 we know that the boundedness of $M : \Lambda_u^p(w) \to \Lambda_u^p(w)$ implies $u(\mathbb{R}^n) = \infty$. This is, essentially, the best we can say about u, since there are examples in which u is not in any A_p class. In fact, the following result proves something stronger: the weight u could be not doubling. See also Proposition 3.4.12.

Theorem 3.3.10 *If $u(x) = e^{|x|}$, $x \in \mathbb{R}$, and $w = \chi_{(0,1)}$, we have that $M : \Lambda_u^q(w) \to \Lambda_u^q(w)$ for every $q > 1$. Therefore, it is not necessary, in general, that the weight u is doubling to have the boundedness $M : \Lambda_u^p(w) \to \Lambda_u^p(w)$.*

Proof. Since the weight w satisfies condition (3.13) of Remark 3.3.6, it is enough to show that, for every cube $Q \subset \mathbb{R}$ we have,

$$\frac{W(u(Q))}{|Q|} \leq C \frac{W(u(E))}{|E|}, \qquad E \subset Q. \tag{3.15}$$

If $u(E) \geq 1$ then $u(Q) \geq 1$, $W(u(E)) = W(u(Q)) = 1$, and (3.15) holds (with $C = 1$) trivially. Thus, we can assume that $u(E) < 1$. Then, $W(u(E)) = u(E)$ and (3.15) is equivalent, by Lebesgue differentiation theorem, to

$$\frac{W(u(Q))}{|Q|} \leq Cu(x), \qquad \text{a.e. } x \in Q. \tag{3.16}$$

To prove (3.16), we can assume $Q = (a,b)$ with $0 \leq a < b$ since u is an even function: for cubes of the form $Q = (-a,b)$ we have

$$\frac{W(u(Q))}{|Q|} \leq \frac{W(2u(0,b))}{|(0,b)|} \leq 2\frac{W(u(0,b))}{|(0,b)|},$$

and we use that the essential infimum of u in $(0, b)$ and in $(-a, b)$ coincide. Let then $Q = (a, b)$ with $0 \leq a < b$ and let us see (3.16). The infimum of u in Q is e^a and, hence, we have to show that $W(u(Q))/|Q| \leq Ce^a$. If $e^b - e^a = u(Q) \geq 1$ then $b \geq \log(1 + e^a)$ and hence

$$\frac{W(u(Q))}{|Q|} = \frac{1}{b-a} \leq \frac{1}{\log(1+e^a) - a} = \frac{1}{\log(1+e^{-a})} \leq Ce^a.$$

If, on the contrary, $e^b - e^a = u(Q) < 1$, we have that $b - a < e^b - e^a < 1$ and therefore,

$$\frac{W(u(Q))}{|Q|} = \frac{u(Q)}{|Q|} = \frac{e^b - e^a}{b - a} \leq e^b = e^a e^{b-a} \leq ee^a. \qquad \square$$

3.4 Weak-type inequality

In this section, we shall study necessary and/or sufficient conditions for the weak-type inequality to hold,

$$M : \Lambda_u^p(w) \longrightarrow \Lambda_u^{p,\infty}(w).$$

The following result (see [CS3]) gives a necessary condition.

Theorem 3.4.1 *Let $0 < p_0, p_1 < \infty$ and let us assume that $M : \Lambda_{u_0}^{p_0}(w_0) \longrightarrow \Lambda_{u_1}^{p_1,\infty}(w_1)$. Then, there exists a constant $C > 0$ such that*

$$\|u_0^{-1}\chi_Q\|_{(\Lambda_{u_0}^{p_0}(w_0))'} \|\chi_Q\|_{\Lambda_{u_1}^{p_1}(w_1)} \leq C|Q| \qquad (3.17)$$

for every cube $Q \subset \mathbb{R}^n$. Here, $\|\cdot\|_{(\Lambda_{u_0}^{p_0}(w_0))'}$ denotes the norm in the associate space (c.f. Definition 2.4.1, Theorem 2.4.7).

In what follows, we shall study condition (3.17) and we shall obtain some important consequences. For example, combining the previous statement with some of the results in chapter 2, we obtain a condition that reduces (depending on w_0) the range of indices p_0 for which the boundedness of $M : \Lambda_{u_0}^{p_0}(w_0) \to \Lambda_{u_1}^{p_1,\infty}(w_1)$ can be true. To this end, we define the index $p_w \in [0, \infty)$ as follows:

$$p_w = \inf\left\{p > 0 : \frac{t^p}{W(t)} \in L^{p'-1}\left((0,1), \frac{dt}{t}\right)\right\}, \qquad (3.18)$$

(where $p' = \infty$ if $0 < p \leq 1$).

Theorem 3.4.2 *Let $0 < p_1 < \infty$ and let us assume that we have the boundedness of $M : \Lambda_{u_0}^{p_0}(w_0) \longrightarrow \Lambda_{u_1}^{p_1,\infty}(w_1)$. Then, $p_0 \geq p_{w_0}$. If $p_{w_0} > 1$ the previous inequality is strict.*

Proof. If $M : \Lambda_{u_0}^{p_0}(w_0) \longrightarrow \Lambda_{u_1}^{p_1,\infty}(w_1)$, we have, by Theorem 3.4.1, $(\Lambda_{u_0}^{p_0}(w_0))' \neq \{0\}$. Then, Theorem 2.4.9 implies $t^{p_0}/W(t) \in L^{p'_0-1}((0,1), dt/t)$. That is,

$$p_0 \in \left\{ p > 0 : \frac{t^p}{W(t)} \in L^{p'-1}\left((0,1), \frac{dt}{t}\right) \right\} = I.$$

The result we are looking for, follows from the fact that I is an interval in $[0, \infty)$, unbounded in the right hand side, and open in the left hand side by p_{w_0}, if $p_{w_0} > 1$. □

Theorem 3.4.3 *If $p_0 < 1$ there are no weights u_0, u_1 such that $M : L^{p_0}(u_0) \to L^{p_1,\infty}(u_1)$, $0 < p_1 < \infty$ is bounded.*

Proof. It is enough to observe that $L^{p_0}(u_0) = \Lambda_{u_0}^{p_0}(1)$ and that, if $w = 1$, $p_w = 1$ (c.f. (3.18)). □

To find equivalent integral expression to (3.17), it will be useful to associate to each weight u in $\mathbb{R}^n$ the family of functions $\{\phi_Q\}_Q$ defined in the following way. For every cube $Q \subset \mathbb{R}^n$,

$$\phi_Q(t) = \phi_{Q,u}(t) = \frac{u(Q)}{|Q|} \int_0^t (u^{-1}\chi_Q)_u^*(s)\, ds, \quad t \geq 0. \qquad (3.19)$$

It will be also very useful the right derivative of the function ϕ_Q

$$\phi'_Q(t) = \frac{u(Q)}{|Q|}(u^{-1}\chi_Q)_u^*(t), \quad t \geq 0. \qquad (3.20)$$

Observe that $\phi'_Q(t) = 0$ if $t \geq u(Q)$, and that $\phi'_Q(t) \leq \phi_Q(t)/t$, $t \geq 0$. Now, we can give an equivalent expression of (3.17) in terms of these functions. These integral expressions are quite similar to those for the classes $B_{p_0,p_1,\infty}$ (Theorem 1.3.3).

Proposition 3.4.4 *Let $\phi_Q = \phi_{Q,u_0}$.*

(a) If $1 < p_0 < \infty$ each of the following expressions are equivalent to (3.17):

(i)
$$\left(\int_0^{u_0(Q)} W_0^{1-p_0'}(s)\, d\phi_Q^{p_0'}(s)\right)^{1/p_0'} W_1^{1/p_1}(u_1(Q)) \leq C u_0(Q).$$

(ii)
$$\left(\int_0^{u_0(Q)} \left(\frac{W_0}{\phi_Q}\right)^{-p_0'}(t) w_0(t)\, dt\right)^{1/p_0'} W_1^{1/p_1}(u_1(Q)) \leq C u_0(Q)$$
$$W_1^{p_0/p_1}(u_1(Q)) \leq C W_0(u_0(Q)).$$

(b) If $0 < p_0 \leq 1$, (3.17) is equivalent to
$$\frac{W_1^{1/p_1}(u_1(Q))}{u_0(Q)} \leq C \frac{W_0^{1/p_0}(t)}{\phi_Q(t)}, \qquad 0 < t \leq u_0(Q).$$

We are assuming both in (a) and (b) that the inequalities are true for every cube $Q \subset \mathbb{R}^n$, and that the constant C is independent of Q.

Proof. Observe that, by definition, $\int_0^t (u_0^{-1}\chi_Q)^*_{u_0}(s)\, ds = \frac{|Q|}{u_0(Q)}\phi_Q(t)$. Hence, by Theorem 2.4.7 (with $r = u_0(Q)$),

$$\begin{aligned}
\|u_0^{-1}\chi_Q\|_{(\Lambda_{u_0}^{p_0}(w_0))'} &\approx \frac{|Q|}{u_0(Q)} \left(\int_0^{u_0(Q)} W_0^{1-p_0'}(s)\, d\phi_Q^{p_0'}(s)\right)^{1/p_0'} \\
&\approx \frac{|Q|}{u_0(Q)} \left(\int_0^{u_0(Q)} \left(\frac{W_0}{\phi_Q}\right)^{-p_0'}(s) w_0(s)\, ds\right)^{1/p_0'} \\
&\quad + \frac{\int_0^\infty (u_0^{-1}\chi_Q)^*_{u_0}(s)\, ds}{W_0^{1/p_0}(u_0(Q))},
\end{aligned}$$

in the case $p_0 > 1$, and

$$\|u_0^{-1}\chi_Q\|_{(\Lambda_{u_0}^{p_0}(w_0))'} = \frac{|Q|}{u_0(Q)} \sup_{t>0} \frac{\phi_Q(t)}{W_0^{1/p_0}(t)},$$

in the case $0 < p_0 \leq 1$. Since $\|\chi_Q\|_{\Lambda_{u_1}^{p_1}(w_1)} = W_1^{1/p_1}(u_1(Q))$ and $\int_0^\infty (u_0^{-1}\chi_Q)^*_{u_0}(s)\, ds = \int_{\mathbb{R}^n} u_0^{-1}(x)\chi_Q(x)u_0(x)\, dx = |Q|$, the given expression can be immediately deduced from (3.17). $\square$

Remark 3.4.5 In many cases, condition (3.17) is also sufficient. For example:

(i) If $w_0 = w_1 = 1$, $p_0 = p_1 = p \geq 1$ inequality (3.17) is

$$\|u_0^{-1}\chi_Q\|_{L^{p'}(u_0)}\|\chi_Q\|_{L^p(u_1)} \leq C|Q|,$$

or equivalently,

$$\left(\frac{1}{|Q|}\int_Q u_1(x)\,dx\right)\left(\frac{1}{|Q|}\int_Q u_0^{-1/(p-1)}(x)\,dx\right)^{p-1} \leq C, \qquad (3.21)$$

in the case $p > 1$, and

$$\frac{u_1(Q)}{|Q|} \leq C\frac{u_0(E)}{|E|}, \qquad E \subset Q,$$

in the case $p = 1$. This is the so-called A_p condition for the pair (u_1, u_0) and it is known to be sufficient for the boundedness $M : L^p(u_0) \to L^{p,\infty}(u_1)$ or, in other terms, $M : \Lambda^p_{u_0}(1) \to \Lambda^{p,\infty}_{u_1}(1)$ (see [GR]).

(ii) If $u_0 = u_1 = 1$, then $\phi_Q(t) = t$, $0 \leq t \leq 1$, and condition (3.17) (or any of the expressions of Proposition 3.4.4) is equivalent to $(w_0, w_1) \in B_{p_0, p_1, \infty}$ (Theorem 1.3.3). That is, the condition is also sufficient.

(iii) If $u_0 = u_1 = u$, $p_0 = p_1 = q \geq 1$ and $w_0(t) = w_1(t) = t^{q/p-1}$, $1 < p < \infty$, we are in the case $M : L^{p,q}(u) \to L^{p,\infty}(u)$. Then condition (3.17) is equivalent to the inequality

$$\|u^{-1}\chi_Q\|_{L^{p',q'}(u)}\|\chi_Q\|_{L^{p,q}(u)} \leq C|Q|. \qquad (3.22)$$

This case was studied by Chung, Hunt, and Kurtz in [CHK] (see also [HK]) where they proved that (3.22) is a necessary and sufficient condition.

The study of the properties of the functions $\{\phi_Q\}$ will allow us to better understand condition (3.17), and obtain some consequences. In the following proposition, we summarize some of these properties.

Proposition 3.4.6 *Let u be a weight in $\mathbb{R}^n$ and, for every cube $Q \subset \mathbb{R}^n$, let $\phi_Q = \phi_{Q,u}$. Then,*

(i) $\phi_Q(t) = \frac{u(Q)}{|Q|} \max\{|E| : E \subset Q, u(E) = t\}, \quad 0 \leq t \leq u(Q).$

(ii) $\phi_{Q|[0,u(Q)]} : [0, u(Q)] \longrightarrow [0, u(Q)]$ *is strictly increasing, onto, concave and absolutely continuous.*

(iii) $\phi_Q(t) \geq t \in [0, u(Q)]$ and $u \in A_1$ if and only if $\phi_Q(t) \leq Ct$, $0 \leq t \leq u(Q)$.

(iv) If $1 < p < \infty$, $0 < q \leq p'$, $u \in A_p \Leftrightarrow \|\phi'_Q(u(Q) \cdot)\|_{L^{p',q}} \leq C$.

(v) If $u \in A_p$, $1 \leq p < \infty$, then $\phi'_Q(u(Q)t) < Ct^{-1/p'}$, $0 < t \leq 1$.

(vi) If $1 < p < \infty$ and $\phi'_Q(u(Q)t) < Ct^{-1/p'}$, $0 < t \leq 1$, then $u \in \bigcap_{q > p} A_q$.

In (iii), (iv), (v), and (vi), C represents a constant not depending on Q.

Proof.
(i) Let $0 \leq t \leq u(Q)$ and $E \subset Q$ be a measurable set such that $u(E) = t$. Then

$$|E| = \int_{\mathbb{R}^n} u^{-1}(x)\chi_Q(x)\chi_E(x)\,u(x)\,dx \leq \int_0^\infty (u^{-1}\chi_Q)^*_u(s)(\chi_E)^*_u(s)\,ds$$
$$= \int_0^t (u^{-1}\chi_Q)^*_u(s)\,ds = \frac{|Q|}{u(Q)}\phi_Q(t).$$

On the other hand, since $(Q, u(x)\chi_Q(x)dx)$ is a strongly resonant measure space (see [BS]), there exists $f \geq 0$ measurable in Q with $f^*_u = (\chi_E)^*_u = \chi_{[0,t)}$ and such that

$$\frac{|Q|}{u(Q)}\phi_Q(t) = \int_0^\infty (u^{-1}\chi_Q)^*_u(s)(\chi_E)^*_u(s)\,ds$$
$$= \int_Q f(x)u^{-1}(x)\chi_Q(x)u(x)\,dx. \qquad (3.23)$$

It follows that $f = \chi_R$, with $R \subset Q$, $u(R) = t$ and, by (3.23), $\frac{|Q|}{u(Q)}\phi_Q(t) = |R|$.

(ii) Let us see that $\phi_{Q|[0,u(Q)]}$ is strictly increasing (the rest of the properties are obvious). It is enough to prove that $(u^{-1}\chi_Q)^*_u(t) > 0$ for $0 < t < u(Q)$. On the contrary, there would exist $t_0 \in (0, u(Q))$ such that $(u^{-1}\chi_Q)^*_u(t_0) = 0$ and hence $u(\{u^{-1}\chi_Q > 0\}) \leq t_0$. Since $t_0 < u(Q)$, we would have that $u(\{x \in Q : u(x) = +\infty\}) = u(\{u^{-1}\chi_Q = 0\}) > 0$, which contradicts the fact that u is locally integrable.

(iii) Since ϕ_Q is concave in $[0, u(Q)]$, $t^{-1}\phi_Q(t)$ is decreasing and, thus, $t^{-1}\phi_Q(t) \geq u(Q)^{-1}\phi_Q(u(Q)) = 1$ for $0 \leq t \leq u(Q)$. The second part of (iii) follows from (i) and the fact that $u \in A_1$ if and only if $\frac{u(Q)}{|Q|} \leq C\frac{u(E)}{|E|}$.

(iv) By definition (see [Stn]), $u \in A_p$

$$\Leftrightarrow \frac{u(Q)}{|Q|}\left(\frac{1}{|Q|}\int_Q u^{-p'/p}(x)\,dx\right)^{p/p'} \leq C$$

$$\Leftrightarrow \frac{u(Q)^{1/p}}{|Q|}\left(\int_{\mathbb{R}^n}(u^{-1}(x)\chi_Q(x))^{p'}u(x)\,dx\right)^{1/p'} \leq C_1$$

$$\Leftrightarrow \frac{u(Q)^{1/p}}{|Q|}\left(\int_0^\infty (u^{-1}\chi_Q)_u^{*p'}(s)\,ds\right)^{1/p'} \leq C_1 \Leftrightarrow \|\phi_Q'(u(Q)\,\cdot\,)\|_{p'} \leq C_1.$$

Since $L^{p',q} \subset L^{p'}$ if $0 < q \leq p'$, we have proved the sufficiency in (iv). Conversely, if $u \in A_p$ it is known that $u \in A_{p-\epsilon}$ for some $\epsilon > 0$. Therefore, by the previous equivalence, $\|\phi_Q'(u(Q)\,\cdot\,)\|_{(p-\epsilon)'} \leq C$. Observe that $\phi_Q'(u(Q)t) = \frac{u(Q)}{|Q|}(u^{-1}\chi_Q)_u^*(u(Q)t) = 0$ for $t \geq 1$. Since $L^{p_1,r}(X) \subset L^{p_0,q}(X)$ if $p_0 < p_1$ and X is of finite measure (see [BS]), we have that

$$\|\phi_Q'(u(Q)\,\cdot\,)\|_{p',q} \leq \|\phi_Q'(u(Q)\,.\,)\|_{(p-\epsilon)'} \leq C.$$

(v) The case $p = 1$ is a consequence of (iii) and the inequality $\phi_Q'(t) \leq \phi_Q(t)/t$. If $u \in A_p$, $1 < p < \infty$ then, by (iv), we have that $\|\phi_Q'(u(Q)\,\cdot\,)\|_{p',\infty} \leq \|\phi_Q'(u(Q)\,\cdot\,)\|_{p'} \leq C$. Since $\phi_Q'(u(Q)\,\cdot\,)$ is decreasing, right continuous and equal to zero in $[1,\infty)$, $\|\phi_Q'(u(Q)\,\cdot\,)\|_{p',\infty} = \sup_{0<t<1} t^{1/p'}\phi_Q'(u(Q)t)$ and we obtain (v).

(vi) The hypothesis implies $\|\phi_Q'(u(Q)\,\cdot\,)\|_{L^{p',\infty}} \leq C$. But $\phi_Q'(u(Q)\,\cdot\,)$ is supported in $[0,1]$ and therefore,

$$\|\phi_Q'(u(Q)\,\cdot\,)\|_{L^{q'}} \leq \|\phi_Q'(u(Q)\,\cdot\,)\|_{L^{p',\infty}} \leq C,$$

for $q > p$ and (vi) follows from (iv). $\square$

Let us see a useful consequence of the two above propositions.

Proposition 3.4.7 *If $0 < p_0, p_1 < \infty$ and $M : \Lambda_{u_0}^{p_0}(w_0) \to \Lambda_{u_1}^{p_1,\infty}(w_1)$, then*

$$(i) \quad \frac{W_1^{1/p_1}(u_1(Q))}{u_0(Q)} \leq C\frac{W_0^{1/p_0}(t)}{\phi_Q(t)}, \quad 0 < t \leq u_0(Q),$$

$$(ii) \quad \frac{W_1^{1/p_1}(u_1(Q))}{u_0(Q)} \leq C\frac{W_0^{1/p_0}(u_0(S))}{u_0(S)}, \quad S \subset Q.$$

Here $\phi_Q = \phi_{Q,u_0}$ and we are assuming that the previous inequalities are satisfied for every cube $Q \subset \mathbb{R}^n$, with C independent of Q.

Proof. (ii) is a consequence of (i) and Proposition 3.4.6 (iii). Hence, we only have to show (i). If $p_0 \leq 1$, (i) is Proposition 3.4.4 (b). For $p_0 > 1$ we have, by Proposition 3.4.4 (a.i),

$$\left(\int_0^t W_0^{1-p_0'}(s)\, d\phi_Q^{p_0'}(s)\right)^{1/p_0'} W_1^{1/p_1}(u_1(Q)) \leq C u_0(Q),$$

for $0 < t \leq u_0(Q)$. Since W_0 is increasing, it follows that

$$W_0^{-1/p_0}(t)\left(\int_0^t d\phi_Q^{p_0'}(s)\right)^{1/p_0'} W_1^{1/p_1}(u_1(Q)) \leq C u_0(Q),$$

and we obtain (i). □

The following result establishes that in the boundedness $M : \Lambda_u^{p_0}(w_0) \to \Lambda_u^{p_1,\infty}(w_1)$ we can always assume that $u \equiv 1$.

Theorem 3.4.8 *If $0 < p_0, p_1 < \infty$ and $M : \Lambda_u^{p_0}(w_0) \to \Lambda_u^{p_1,\infty}(w_1)$, we have that*

$$M : \Lambda^{p_0}(w_0) \longrightarrow \Lambda^{p_1,\infty}(w_1),$$

that is, $(w_0, w_1) \in B_{p_0,p_1,\infty}$.

Proof. If $p_0 > 1$ then, by Proposition 3.4.4 (a.ii) and using that $\phi_Q(t) \geq t$, $0 < t < u(Q)$ (Proposition 3.4.6 (iii)) we have the inequalities

$$\left(\int_0^{u(Q)} \left(\frac{W_0(t)}{t}\right)^{-p_0'} w_0(t)\, dt\right)^{1/p_0'} W_1^{1/p_1}(u(Q)) \leq C u(Q),$$
$$W_1^{1/p_1}(u(Q)) \leq C W_0^{1/p_0}(u(Q)),$$

for every cube $Q \subset \mathbb{R}^n$. But we have proved (Proposition 3.2.7) that $u(\mathbb{R}^n) = \infty$. Then, for every $r > 0$ there exists a cube Q with $u(Q) = r$. Therefore, the two previous inequalities are equivalent to the condition (a.iii) of Theorem 1.3.3 and hence, $(w_0, w_1) \in B_{p_0,p_1,\infty}$.

If $p_0 \leq 1$ we apply Proposition 3.4.4(b) to obtain the expression of Theorem 1.3.3 to conclude the same. □

Corollary 3.4.9 $B_{p,\infty}(u) \subset B_{p,\infty}$, $0 < p < \infty$.

Remark 3.4.10 (i) If $M : \Lambda_u^1(w) \to \Lambda_u^{1,\infty}(w)$ then, by Corollary 3.4.9, $w \in B_{1,\infty}$. The bigger class B_p contained in $B_{1,\infty}$ is B_1 and if $w \in B_1$, Theorem 3.2.11 characterizes the previous boundedness. That is, the problem $M : \Lambda_u^1(w) \to \Lambda_u^{1,\infty}(w)$ only remains open in the case $w \in B_{1,\infty} \setminus B_1$.

(ii) Condition (3.17) is sufficient in the case $M : \Lambda_u^1(w) \to \Lambda_u^{1,\infty}(w)$ if the weight w satisfies

$$\sum_{j=1}^n \frac{W(t_j)}{W(r_j)} r_j \leq C \frac{W(\sum_j t_j)}{W(\sum_j r_j)} \sum_j r_j,$$

for every finite families of numbers $\{(t_j, r_j)\}_j$ with $0 < t_j < r_j$, $j = 1, \ldots, n$. To see this, we observe that if the previous condition and (3.17) hold (which implies (3.7)) we have, for every finite family of disjoint cubes $(Q_j)_j$ and every family of sets $(E_j)_j$ with $E_j \subset Q_j$,

$$\sum_j \frac{|E_j|}{|Q_j|} u(Q_j) \leq C \sum_j \frac{W(u(E_j))}{W(u(Q_j))} u(Q_j) \leq C \frac{W(u(\bigcup E_j))}{W(u(\bigcup Q_j))} \sum_j u(Q_j).$$

This also holds (since $W \in \Delta_2$) if the cubes Q_j are "almost" disjoint (if $\sum_j \chi_{Q_j} \leq k = k_n$). If $0 \leq f \in \mathcal{M}(\mathbb{R}^n)$ is bounded and with compact support and $t > 0$, the level set $E_t = \{Mf > t\}$ is contained (see [LN1]) in a finite union of cubes $(Q_j)_j$ such that $\sum_j \chi_{Q_j} \leq k = k_n$ (only depending on the dimension) and $\frac{1}{|Q_j|} \int_{Q_j} f(x)\, dx > t$, for every j. Then,

$$\begin{aligned} t \sum_j u(Q_j) &\leq \int_{\mathbb{R}^n} f(x) \Big(\sum_j \frac{u(Q_j)}{|Q_j|} u^{-1}(x) \chi_{Q_j}(x) \Big) u(x)\, dx \\ &\leq \|f\|_{\Lambda_u^1(w)} \Big\| \sum_j \frac{u(Q_j)}{|Q_j|} u^{-1} \chi_{Q_j} \Big\|_{(\Lambda_u^1(w))'} \end{aligned}$$

and, to prove that $\|Mf\|_{\Lambda_u^{1,\infty}(w)} \leq C \|f\|_{\Lambda_u^1(w)}$ is bounded, we only have to see that

$$\Big\| \sum_j \frac{u(Q_j)}{|Q_j|} u^{-1} \chi_{Q_j} \Big\|_{(\Lambda_u^1(w))'} \lesssim \frac{\sum_j u(Q_j)}{W(u(\bigcup_j Q_j))},$$

(since then we have that $tW(u(E_t)) \leq tW(u(\bigcup_j Q_j)) \leq C\|f\|_{\Lambda_u^1(w)}$). But, since $\sum_j \chi_{Q_j} \leq k$,

$$\int_0^t \Big(\sum_j \frac{u(Q_j)}{|Q_j|} u^{-1} \chi_{Q_j} \Big)_u^*(s)\, ds = \sup_{u(E)=t} \int_E \sum_j \frac{u(Q_j)}{|Q_j|} u^{-1}(x) \chi_{Q_j}(x)\, u(x)\, dx$$

$$= \sup_{u(E)=t} \sum_j \frac{|E \cap Q_j|}{|Q_j|} u(Q_j)$$

$$\lesssim \frac{W(t)}{W(u(\bigcup_j Q_j))} \sum_j u(Q_j),$$

and (see Theorem 2.4.7(i)) the inequality we are looking for follows.

(iii) If $w \in L^1(\mathbb{R}^+)$, the condition on w of the previous observation can be weakened up: it is sufficient that it holds "near 0", that is, with $\sum r_j < \epsilon$ (for some fixed $\epsilon > 0$).

(iv) It is easy to see (using the two previous observations) that, for the weights $w = \chi_{(0,1)}$, $w(t) = (\log^+(1/t))^\alpha$, $w(t) = (\log^+ \log^+(1/t))^\alpha$, $w(t) = t^\alpha$, condition (3.17) is also equivalent (in the case $p_0 = p_1 = 1$, $u_0 = u_1$, $w_0 = w_1 = w$.)

Theorem 3.4.11 *Let $1 \leq p < \infty$, $0 < q < \infty$. Then,*

(i) $B_{q/p} \subset B_{q,\infty}(u)$ implies $u \in \bigcap_{r>p} A_r$.
(ii) If $q \leq 1$, $B_{q,\infty} \subset B_{q,\infty}(u)$ implies $u \in A_1$.

Proof. (i) It is immediate to check (Theorem 1.3.4) that the weight $w(t) = t^{r-1}$ is in $B_{q/p} \subset B_{q,\infty}(u)$, for $0 < r < q/p$. Hence, by Proposition 3.4.7 (i), we have that, for every cube $Q \subset \mathbb{R}^n$ and $0 < t < u(Q)$,

$$\phi_Q(t) \leq Cu(Q) \frac{W^{1/q}(t)}{W^{1/q}(u(Q))} = Cu(Q)^{1-r/q} t^{r/q}.$$

Since $\phi'_Q(t) \leq \phi_Q(t)/t$, $0 < t < u(Q)$, we obtain that $\phi'_Q(u(Q)t) \leq Ct^{r/q-1}$, $0 < t < 1$, and by Proposition 3.4.6(vi), $u \in A_s$ for every $s > q/r$. Since this holds for $0 < r < q/p$, we conclude that $u \in \bigcap_{r>p} A_r$.

(ii) Observe that (using, for example, Theorem 1.3.3) $w(t) = t^{q-1} \in B_{q,\infty} \subset B_{q,\infty}(u)$. From Proposition 3.4.7(i) (with $w_0 = w_1 = t^{q-1}$, $u_0 = u_1 = u$, $p_0 = p_1 = q$) we obtain now that $\phi_Q(t) \leq Ct$, $0 < t < u(Q)$, and the result is a consequence of Proposition 3.4.6 (iii). □

The following result is related to the A_∞ class. Let us recall that (see, for example, [Stn]) $u \in A_\infty = \bigcup_{p \geq 1} A_p$ if and only if there exist constants $\epsilon, C > 0$ so that

$$\frac{u(Q)}{|Q|^\epsilon} \geq C \frac{u(E)}{|E|^\epsilon} \tag{3.24}$$

for every cube Q and every measurable set $E \subset Q$. In general, the boundedness of $M : \Lambda_u^p(w) \to \Lambda_u^{p,\infty}(w)$ does not imply $u \in A_\infty$ (Theorem 3.3.10). But we can still give a sufficient condition on the weight w.

Proposition 3.4.12 *Let us assume that there exist constants $\epsilon, C > 0$ so that*
$$\frac{W(r)}{r^\epsilon} \geq C \frac{W(t)}{t^\epsilon}, \quad 0 < t < r < \infty.$$
Then $A_p(w) \subset A_\infty$, $0 < p < \infty$.

Proof. We can assume $\epsilon < p$. If $u \in A_p(w)$, by Proposition 3.4.7,
$$\phi_Q(t) \leq C_1 \frac{W^{1/p}(t)}{W^{1/p}(u(Q))} u(Q) \leq C_2 \left(\frac{t}{u(Q)} \right)^{\epsilon/p} u(Q),$$
for every Q and $0 < t < u(Q)$. Then,
$$\phi'_Q(u(Q)t) \leq \frac{1}{u(Q)t} \phi_Q(u(Q)t) \leq C_2 t^{\epsilon/p - 1}, \quad 0 < t < 1,$$
and, from Proposition 3.4.6 (vi), it follows that $u \in A_{q'}$, for the range $q < p/(p - \epsilon)$. $\square$

Remark 3.4.13 The condition on w of Proposition 3.4.12 is not equivalent to $w \in \bigcup_p B_p$ (as it happens with the classes A_p). For example, the weight $w = \chi_{(0,1)} \in B_{1,\infty}$ does not satisfy that condition. Even if we impose $w \notin L^1(\mathbb{R}^+)$ as the example $w(t) = 1/(1+t)$ shows.

Power weights $w(t) = t^\alpha$ trivially satisfy the condition of Proposition 3.4.12. Another nontrivial example, of a function satisfying such condition (with $\epsilon = 1/2$) is $w(t) = 1 + \log^+(1/t)$.

Up to now, we have only seen necessary conditions for the weak-type boundedness. Let us see now some sufficient conditions. For example, from Theorem 3.2.2, we deduce the two following results.

Theorem 3.4.14 *Let $0 < p, q < \infty$ and let us assume that $M : \Lambda_u^p(w) \to \Lambda_u^{p,\infty}(w)$ is bounded. Let $\tilde{w}$ another weight. Then we have that $M : \Lambda_u^q(\tilde{w}) \to \Lambda_u^{q,\infty}(\tilde{w})$ is bounded in each of the following cases:*

(i) If $0 < q \leq p$ and the condition

$$\frac{\widetilde{W}^{1/q}(r)}{W^{1/p}(r)} \leq C \frac{\widetilde{W}^{1/q}(t)}{W^{1/p}(t)}, \qquad 0 < t < r < \infty$$

holds.

(ii) If $p < q < \infty$ and it satisfies

$$\left(\frac{1}{\widetilde{W}(t)} \int_0^t \left(\frac{W(s)}{\widetilde{W}(s)}\right)^{\frac{q}{q-p}} \tilde{w}(s)\, ds\right)^{\frac{q-p}{q}} \leq C \frac{W(t)}{\widetilde{W}(t)}, \qquad t > 0.$$

Proof. Let us fix $t > 0$ and let us consider the weights

$$w_1(s) = \frac{w(s)}{W(t)} \chi_{(0,t)}(s), \quad w_0(s) = \frac{\tilde{w}(s)}{\widetilde{W}(t)} \chi_{(0,t)}(s), \quad s > 0.$$

Under the hypothesis of (i) we have, by Theorem 1.2.18 (d)

$$\sup_{g\downarrow} \frac{\left(\int_0^\infty g^p(t) w_1(t)\, dt\right)^{1/p}}{\left(\int_0^\infty g^q(t) w_0(t)\, dt\right)^{1/q}} = \sup_{s>0} \frac{W_1^{1/p}(s)}{W_0^{1/q}(s)} \leq C.$$

Therefore,

$$\left(\frac{1}{W(t)} \int_0^t (f_u^*)^p(s) w(s)\, ds\right)^{1/p} \leq C \left(\frac{1}{\widetilde{W}(t)} \int_0^t (f_u^*)^q(s) \tilde{w}(s)\, ds\right)^{1/q}$$

and we obtain (i).

To see (ii), observe that the hypothesis implies, with $r = q/p > 1$,

$$B = B(t) = \left(\int_0^\infty \left(\frac{W_1(s)}{W_0(s)}\right)^{r'} w_0(s)\, ds\right)^{1/r'} + \frac{W_1(\infty)}{W_0^{1/r}(\infty)} \leq C + 1 < \infty,$$

and by Theorem I.5.7 in [Sa], we have

$$\sup_{g\downarrow} \frac{\left(\int_0^\infty g^p(s) w_1(s)\, ds\right)^{1/p}}{\left(\int_0^\infty g^q(s) w_0(s)\, ds\right)^{1/q}} = \left(\sup_{g\downarrow} \frac{\int_0^\infty g(s) w_1(s)\, ds}{\left(\int_0^\infty g^r(s) w_0(s)\, ds\right)^{1/r}}\right)^{1/p}$$

$$\leq (C'B)^{1/p} \leq (C'(C+1))^{1/p} = C'' < \infty.$$

In particular,

$$\left(\frac{1}{W(t)}\int_0^t (f_u^*)^p(s)w(s)\,ds\right)^{1/p} \leq C''\left(\frac{1}{\widetilde{W}(t)}\int_0^t (f_u^*)^q(s)\widetilde{w}(s)\,ds\right)^{1/q},$$

for every $t > 0$ and $f \in \mathcal{M}(\mathbb{R}^n)$ and by Theorem 3.2.2 we get the result. $\square$

Corollary 3.4.15 *Let $0 < p < \infty$ and let us assume that $M : \Lambda_u^p(w) \to \Lambda_u^{p,\infty}(w)$ is bounded. If $0 < q \leq p$ and $\widetilde{w}$ is a weight such that $\widetilde{W}^{1/q}/W^{1/p}$ is decreasing, then $M : \Lambda_u^q(\widetilde{w}) \longrightarrow \Lambda_u^{q,\infty}(\widetilde{w})$ is also bounded.*

The following result completes Theorem 3.4.8.

Theorem 3.4.16 *If $u \in A_1$, we have the boundedness $M : \Lambda_u^{p_0}(w_0) \to \Lambda_u^{p_1,\infty}(w_1)$, if and only if $M : \Lambda^{p_0}(w_0) \to \Lambda^{p_1,\infty}(w_1)$. In particular $B_{p,\infty}(u) = B_{p,\infty}$, $0 < p < \infty$, if $u \in A_1$.*

Proof. The "only if" condition is Theorem 3.4.8. To see the other implication, let us observe that if $u \in A_1$ we have, by Theorem 3.2.2 (applied to $w \equiv 1$, $p = 1$) that $(Mf)_u^*(t) \lesssim Af_u^*(t)$, $t > 0$, $f \in \mathcal{M}(\mathbb{R}^n)$, where A the Hardy operator. Since the boundedness of $M : \Lambda^{p_0}(w_0) \to \Lambda^{p_1,\infty}(w_1)$ is equivalent to $A : L_{\mathrm{dec}}^{p_0}(w_0) \to L^{p_1,\infty}(w_1)$, we get

$$W_1^{1/p_1}(t)(Mf)_u^*(t) \lesssim \|Af_u^*\|_{L^{p_1,\infty}(w_1)} \lesssim \|f_u^*\|_{L^{p_0}(w_0)} = \|f\|_{\Lambda_u^{p_0}(w_0)}, \qquad t > 0,$$

which is equivalent to $\|Mf\|_{\Lambda_u^{p_1,\infty}(w_1)} \lesssim \|f\|_{\Lambda_u^{p_0}(w_0)}$. $\square$

Using the same idea, we can prove the following result analogous to Theorem 3.2 in [CS3].

Theorem 3.4.17 *Let $0 < p_0, p_1 < \infty$, $1 \leq p < \infty$ and let us assume $u \in A_p$. Then,*

(i) *If $(w_0, w_1) \in B_{p_0,p_1}$, we have that $M : \Lambda_u^{pp_0}(w_0) \to \Lambda_u^{pp_1}(w_1)$ is bounded.*

(ii) *If $(w_0, w_1) \in B_{p_0,p_1,\infty}$, then $M : \Lambda_u^{pp_0}(w_0) \to \Lambda_u^{pp_1,\infty}(w_1)$ is bounded.*

Proof. If $u \in A_p$, by Theorem 3.2.2, $(Mf)_u^*(t)^p \lesssim A((f_u^*)^p)(t)$. The hypothesis in (i) implies that $A: L^{p_0}_{\text{dec}}(w_0) \to L^{p_1}(w_1)$ and it follows that

$$\|Mf\|^p_{\Lambda^{pp_1}_u(w_1)} = \left(\int_0^\infty ((Mf)_u^*)^{pp_1}(t) w_1(t)\, dt\right)^{1/p_1} \lesssim \|A((f_u^*)^p)\|_{L^{p_1}(w_1)}$$
$$\lesssim \|(f_u^*)^p\|_{L^{p_0}(w_0)} = \|f\|^p_{\Lambda^{pp_0}_u(w_0)}.$$

Analogously, one can easily prove (ii). □

If $p > 1$, $p_0 \leq p_1$, (ii) can be improved in the following way:

Corollary 3.4.18 *Let $0 < p_0 \leq p_1 < \infty$. If $u \in A_p$, $p > 1$, and $(w_0, w_1) \in B_{p_0, p_1, \infty}$, then $M: \Lambda^{pp_0}_u(w_0) \to \Lambda^{pp_1}_u(w_1)$ is bounded.*

Proof. If $u \in A_p$ then $u \in A_q$ for some $q \in (1,p)$, and by the previous theorem, $M: \Lambda^{qp_0}_u(w_0) \to \Lambda^{qp_1,\infty}_u(w_1)$ is bounded. Applying now the interpolation theorem (Theorem 2.6.3) we obtain the boundedness of $M: \Lambda^{pp_0}_u(w_0) \to \Lambda^{pp_1,pp_0}_u(w_1) \subset \Lambda^{pp_1}_u(w_1)$. □

Remark 3.4.19 In [Ne2], Neugebauer introduces the classes A_p^* which are defined by

$$A_p^* = \{u \in A_\infty : p = \inf\{q \geq 1 : u \in A_q\}\}, \qquad 1 \leq p < \infty.$$

These classes are pairwise disjoint and the union of all of them is A_∞. With this terminology, Theorem 3.3.7 states that $B_p(u) = B_p$, $0 < p < \infty$, if $u \in A_1^*$ (and this condition is also necessary). From Theorems 3.4.11 and 3.4.17 we can deduce (using the property $w \in B_p \Rightarrow w \in B_{p-\epsilon}$) that, for $1 < p < \infty$, $0 < q < \infty$,

(a) $u \in A_p^* \Rightarrow B_{q/p} \subset B_q(u)$,
(b) $B_{q/p} = B_q(u) \Rightarrow u \in A_p^*$.

This cannot be improved (as it happens in the case $p = 1$), that is, it is not true that the characterization of the classes $B_q(u)$ when $u \in A_\infty$, is $B_q(u) = B_{p/q}$ for p such that $u \in A_p^*$. In fact, it is easy to check that the weight $u(x) = 1 + |x|$, $x \in \mathbb{R}$, is in A_2^* and, by Theorem 3.3.5 (see also Remark 3.3.6(ii)), $w = \chi_{(0,1)} \in B_{3/2}(u)$ (condition (3.12) holds with $q = 1$). However, $w \notin B_{3/4}$ and thus, $B_{3/2}(u) \neq B_{3/4}$.

We shall study now a sufficient condition to have the weak-type boundedness. To this end, we shall need the following notation. We shall associate to each weight u in $\mathbb{R}^n$ a function Φ_u which is connected with the family of functions $\{\phi_Q\}_Q$ introduced in (3.19). This new function is defined by

$$\Phi_u(t) = \sup_Q \phi'_Q(u(Q)\,t), \quad 0 < t < \infty, \tag{3.25}$$

where the supremum is taken over all cubes $Q \subset \mathbb{R}^n$. Let us observe that

$$\Phi_u(t) = \sup_Q \frac{u(Q)}{|Q|}(u^{-1}\chi_Q)^*_u(u(Q)\,t), \quad 0 < t < \infty.$$

Therefore, $\Phi_u(t) = 0$ if $t \geq 1$. Moreover, by Proposition 3.4.6(iii), $\int_0^t \Phi_u(s)\,ds \geq t$, $0 < t < 1$. We know that if $u \equiv 1$, $(Mf)^*_u(t) \approx A(f^*_u)(t)$, $t > 0$, where A is the Hardy operator. And although this does not happen when $u \neq 1$ (see [CS3]) there exists the following positive partial result due to Leckband and Neugebauer ([LN1]).

Theorem 3.4.20 *Let u be a weight in $\mathbb{R}^n$. Then, for each $f \in \mathcal{M}(\mathbb{R}^n)$ we have that*

$$(Mf)^*_u(t) \leq C \int_0^\infty \Phi_u(s) f^*_u(st)\,ds, \quad 0 < s < \infty,$$

where C is a constant depending only on the dimension.

Using this theorem, we can find a sufficient condition to have the boundedness $M : \Lambda^{p_0}_u(w_0) \to \Lambda^{p_1,\infty}_u(w_1)$. Observe the analogy with the corresponding expressions of Proposition 3.4.4.

Theorem 3.4.21 *Let $0 < p_1 < \infty$.*

(a) If $1 < p_0 < \infty$ and there exists a constant $C < \infty$ such that

(i) $\left(\int_0^r \left(\int_0^{t/r} \Phi_u(s)\,ds \right)^{p'_0} W_0^{-p'_0}(t) w_0(t)\,dt \right)^{1/p'_0} W_1^{1/p_1}(r) \leq C, \quad r > 0,$

(ii) $W_1^{1/p_1}(r) \leq C W_0^{1/p_0}(r), \quad r > 0,$

then $M : \Lambda^{p_0}_u(w_0) \to \Lambda^{p_1,\infty}_u(w_1)$ is bounded.

(b) If $0 < p_0 \leq 1$, we have the same if the following condition holds

$$\int_0^{t/r} \Phi_u(s)\,ds \leq C \frac{W_0^{1/p_0}(t)}{W_1^{1/p_1}(r)}, \quad 0 < t < r < \infty.$$

Proof. Let us consider the operator

$$Tg(t) = \int_0^\infty \Phi_u(s) g(st)\, ds = \int_0^\infty k(t,s) g(s)\, ds,$$

(with $k(t,s) = (1/t)\Phi_u(s/t)$) acting on functions $g \downarrow$. By Theorem 3.4.20, the boundedness of

$$T : L^{p_0}_{\text{dec}}(w_0) \longrightarrow L^{p_1,\infty}(w_1) \tag{3.26}$$

implies the result. The characterization of (3.26) can be obtained as a direct application of Theorem 4.3 in [CS2]. □

Remark 3.4.22 (i) The sufficient condition of the previous theorem is not necessary in general. For example, if $w_0 = w_1 = w = \chi_{(0,1)}$, $p_0 = p_1 = 1$, such condition is $\int_0^t \Phi_u(s)\, ds \lesssim t$, $0 < t < 1$, and by Proposition 3.4.6, it is equivalent to $u \in A_1$. That is, $A_1 \subset A_1(\chi_{(0,1)})$. However, for this weight, condition (3.17) is necessary and sufficient in this case (see Remark 3.4.10). This condition is (Proposition 3.4.4b) $\phi_Q(t) \lesssim \dfrac{u(Q)}{W(u(Q))} W^{-1}(t)$, $t > 0$, and, by Proposition 3.4.6(i), it is equivalent to

$$\frac{W(u(Q))}{|Q|} \leq C \frac{W(u(E))}{|E|}, \qquad E \subset Q.$$

It is a simple exercise to check that (with $w = \chi_{(0,1)}$) the weight $u(x) = 1 + |x|$, $x \in \mathbb{R}$, satisfies this condition. However, $u \notin A_p$ if $p \leq 2$, and it follows that $A_1(\chi_{(0,1)}) \neq A_1$, that is, the condition of Theorem 3.4.21 is not necessary.

(ii) Let us assume that the weight u satisfies the following property: for every $r > 0$ there exists a cube Q_r so that

(a) $u(Q_r) \approx r$,

(b) $\int_0^t \Phi_u(s)\, ds \approx \int_0^t \phi_{Q_r}(u(Q_r)s)\, ds$, $0 < t < 1$.

Then the condition of the previous theorem is equivalent to (3.17) and, hence, both are equivalent to the boundedness of $M : \Lambda^{p_0}_u(w_0) \to \Lambda^{p_1,\infty}_u(w_1)$. This follows immediately from the expressions of Theorem 3.4.21 and by Proposition 3.4.4 (with $u_0 = u_1 = u$) since, fixed $r > 0$ one can substitute, in the expressions of Theorem 3.4.21, r by $u(Q_r)$ and the integral $\int_0^{t/r} \Phi_u(s)\, ds$

by $\int_0^{t/r} \phi_Q(u(Q_r)s)\,ds$ to obtain the condition of Proposition 3.4.4 and viceversa.

(iii) Every power weight $u(x) = |x|^\alpha$, $x \in \mathbb{R}^n$, $\alpha > 0$ satisfies the previous condition. In fact, if Q is a cube centered at the origin, one can easily see that

$$\phi_Q(u(Q)t) \approx \Phi_u(t) \approx t^{\frac{-\alpha}{n+\alpha}}, \qquad 0 < t < 1.$$

Therefore, the characterization of the boundedness of $M : \Lambda_u^{p_0}(w_0) \to \Lambda_u^{p_1,\infty}(w_1)$ is given in this case by the expressions of Theorem 3.4.21 (which is now equivalent to (3.17)). See also Theorem 5.7.

3.5 Applications

We can use now the result of the previous sections, to characterize in its total generality, the boundedness of

$$M : L^{p,q}(u) \to L^{r,s}(u), \qquad (3.27)$$

completing results of [Mu, CHK, HK, La].

Theorem 3.5.1 *Let $p, r \in (0, \infty)$, $q, s \in (0, \infty]$.*

(a) *If either $p < 1$, $p \neq r$ or $s < q$, there are no weights u satisfying the boundedness $M : L^{p,q}(u) \to L^{r,s}(u)$.*

(b) *The boundedness of*

$$M : L^{1,q}(u) \longrightarrow L^{1,s}(u)$$

only holds if $q \leq 1$, $s = \infty$ and, in this case, a necessary and sufficient condition is $u \in A_1$.

(c) *If $p > 1$ and $0 < q \leq s \leq \infty$, a necessary and sufficient condition to have the boundedness of*

$$M : L^{p,q}(u) \longrightarrow L^{p,s}(u)$$

is:

(1) *If $q \leq 1$, $s = \infty$:* $\quad \dfrac{u(Q)}{|Q|^p} \leq C \dfrac{u(E)}{|E|^p}, \quad E \subset Q.$

(2) If $q > 1$ or $s < \infty$: $u \in A_p$.

Proof. (a) By Proposition 3.2.9, the boundedness of

$$M : L^{p,q}(u) \longrightarrow L^{r,s}(u) \tag{3.28}$$

implies $s \geq q$. On the other hand, (3.28) implies the boundedness of $M : L^{p,q}(u) \to L^{r,\infty}(u)$ which is equivalent to $M : \Lambda_u^q(t^{q/p-1}) \to \Lambda_u^{r,\infty}(1)$ and from Corollary 3.2.6 it follows that

$$\frac{(u(Q))^{1/r}}{|Q|} \leq C \frac{(u(E))^{1/p}}{|E|}, \quad E \subset Q. \tag{3.29}$$

By the Lebesgue differentiation theorem, we have that necessarily $p \geq 1$. Moreover, applying (3.29) with $E = Q$ we obtain that $(u(Q))^{1/r-1/p} \leq C$ and since $u(Q)$ can take any value in $(0, \infty)$ (Proposition 3.2.7), we get $p = r$.

(b) Since the norm of a characteristic function in $L^{1,q}$ does not depend on q, the boundedness in (b) implies $\|M\chi_E\|_{L^{1,s}(u)} \leq C\|\chi_E\|_{L^1(u)}$, $E \subset Q$. If $s < \infty$ we obtain, from Theorem 3.3.3, the boundedness of $M : L^p(u) \to L^p(u)$ with $p < 1$ and, by (a), we have a contradiction. Therefore, $s = \infty$. On the other hand, the boundedness of $M : L^{1,q}(u) \to L^{1,\infty}(u)$ is the same than $M : \Lambda_u^q(t^{q-1}) \to \Lambda_u^{q,\infty}(t^{q-1})$ and by Corollary 3.2.6 we obtain $u(Q)/|Q| \leq Cu(E)/|E|$, $E \subset Q$, which is $u \in A_1$. We know that this condition is sufficient if $q \leq 1$. But from Corollary 3.4.9 we get $t^{q-1} \in B_{q,\infty}$, and this is only possible (see Theorem 1.3.3) if $q \leq 1$.

(c) If $q \leq s < \infty$, by Theorem 3.3.3 and using interpolation, the boundedness of $M : L^{p,q}(u) \to L^{p,s}(u)$ is equivalent to the boundedness of $M : L^p(u) \to L^p(u)$ and a necessary and sufficient condition is $u \in A_p$. In the case $M : L^{p,q}(u) \to L^{p,\infty}(u)$ (that is $s = \infty$) we have two possibilities: (i) if $q > 1$ a necessary and sufficient condition is (see [CHK]) $u \in A_p$, and (ii) if $q \leq 1$, from Corollary 3.2.6 we obtain the condition $u(Q)/|Q|^p \leq u(E)/|E|^p$. In [CHK] it is shown that this condition is sufficient in the case $q = 1$ and (since $L^{p,q} \subset L^{p,1}$ if $q \leq 1$) also in the case $q \leq 1$. □

Remark 3.5.2 (i) The cases $M : L^1(u) \to L^{1,\infty}(u)$ and $M : L^{p,q}(u) \to L^{p,s}(u)$ with $1 < p \leq q \leq s \leq \infty$ were solved by Muckenhoupt in [Mu], giving rise to the A_p classes. Chung, Hunt, and Kurtz ([CHK, HK]) solved the case $M : L^{p,q}(u) \to L^{p,s}(u)$ with $p > 1$, $1 \leq q \leq s \leq \infty$ and Lai ([La]) proved the necessity of the condition $p = r$ to have (3.27).

(ii) The conditions obtained in the cases of weak-type inequalities ($q, s < \infty$) coincide with (3.17).

The following result characterizes the boundedness of

$$M : \Lambda_u^{p_0}(w_0) \to \Lambda_u^{p_1,\infty}(w_1) \qquad (3.30)$$

when $u(x) = |x|^\alpha$, $x \in \mathbb{R}^n$.

Theorem 3.5.3 *Let $u(x) = |x|^\alpha$, $x \in \mathbb{R}^n$, $\alpha > -n$.*

(a) If $\alpha \leq 0$, (3.30) is equivalent to $(w_0, w_1) \in B_{p_0,p_1,\infty}$.

(b) If $\alpha > 0$ (3.30) holds if and only if $(\bar{w}_0, \bar{w}_1) \in B_{p_0,p_1,\infty}$, where for each $i = 0, 1$,

$$\bar{w}_i(t) = w_i(t^{\frac{n+\alpha}{n}}) t^{\frac{\alpha}{n}}, \qquad t > 0.$$

Proof. (a) is consequence of Theorem 3.4.16, since $u(x) = |x|^\alpha \in A_1$ if $\alpha \leq 0$. To prove (b) we use the condition of Theorem 3.4.21 that (by Remark 3.4.22(iii)) is necessary and sufficient. The final condition is obtained making the change of variables $t = \bar{t}^{(n+\alpha)/n}$, $r = \bar{r}^{(n+\alpha)/n}$ and comparing the expression we get with the corresponding to the classes $B_{p_0,p_1,\infty}$ (Theorem 1.3.3). □

Bibliography

[Al]   G.D. Allen, *Duals of Lorentz spaces*, Pacific J. Math. **177** (1978), 287–291.

[AEP] M.A. Ariño, R. Eldeeb, and N.T. Peck, *The Lorentz sequence spaces $d(w,p)$ where w is increasing*, Math. Ann. **282** (1988), 259–266.

[AM1] M.A. Ariño and B. Muckenhoupt, *Maximal functions on classical Lorentz spaces and Hardy's inequality with weights for nonincreasing functions*, Trans. Amer. Math. Soc. **320** (1990), 727–735.

[AM2] M.A. Ariño and B. Muckenhoupt, *A characterization of the dual of the classical Lorentz sequence space $d(w,q)$*, Proc. Amer. Math. Soc. **102** (1991), 87–89.

[BPS] S. Barza, L.E. Persson, and J. Soria, *Sharp weighted multidimensional integral inequalities for monotone functions*, Math. Nachr. **210** (2000), 43–58.

[BS]   C. Bennett and R. Sharpley, *Interpolation of Operators*, Academic Press, 1988.

[BL]   J. Bergh and J. Löfström, *Interpolation Spaces. An Introduction*, Springer-Verlag, New York, 1976.

[CH]   J.R. Calder and J.B. Hill, *A collection of sequence spaces*, Trans. Amer. Math. Soc. **152** (1970), 107–118.

[CGS] M.J. Carro, A. García del Amo, and J. Soria, *Weak-type weights and normable Lorentz spaces*, Proc. Amer. Math. Soc. **124** (1996), 849–857.

[CPSS] M.J. Carro, L. Pick, J. Soria, and V. Stepanov, *On embeddings between classical Lorentz spaces*, Math. Inequal. Appl. **4** (2001), 397-428.

[CS1] M.J. Carro and J. Soria, *Weighted Lorentz spaces and the Hardy operator*, J. Funct. Anal. **112** (1993), 480–494.

[CS2] M.J. Carro and J. Soria, *Boundedness of some integral operators*, Canad. J. Math. **45** (1993), 1155–1166.

[CS3] M.J. Carro and J. Soria, *The Hardy-Littlewood maximal function and weighted Lorentz spaces*, J. London Math. Soc. **55** (1997), 146–158.

[CM] J. Cerdà and J. Martín, *Interpolation restricted to decreasing functions and Lorentz spaces*, Proc. Edinburgh Math. Soc. **42** (1999), 243–256.

[CHK] H.M. Chung, R.A. Hunt, and D.S. Kurtz, *The Hardy-Littlewood maximal function on $L(p,q)$ spaces with weights*, Indiana Univ. Math. J. **31** (1982), 109–120.

[CF] R.R. Coifman and C. Fefferman, *Weighted norm inequalities for maximal functions and singular integrals*, Studia Math. **51** (1974), 241–250.

[Cw1] M. Cwikel, *On the conjugates of some function spaces*, Studia Math. **45** (1973), 49–55.

[Cw2] M. Cwikel, *The dual of weak L^p*, Ann. Inst. Fourier (Grenoble) **25** (1975), 81–126.

[CS] M. Cwikel and Y. Sagher, *$L(p,\infty)^*$*, Indiana Univ. Math. J. **21** (1972), 781–786.

[GR] J. García Cuerva and J.L. Rubio de Francia, *Weighted Norm Inequalities and Related Topics*, vol. 46, North-Holland, 1985.

[Ga] A.J.H. Garling, *A class of reflexive symmetric BK-spaces*, Canad. J. Math. **21** (1969), 602–608.

[Ha]   A. Haaker, *On the conjugate space of Lorentz space*, University of Lund (1970).

[HM]   H. Heinig and L. Maligranda, *Weighted inequalities for monotone and concave functions*, Studia Math. **116** (1995), 133–165.

[Hu]   R.A. Hunt, *On $L(p,q)$ spaces*, Enseignement Math. **12** (1966), 249–276.

[HK]   R.A. Hunt and D.S. Kurtz, *The Hardy-Littlewood maximal function on $L(p,1)$*, Indiana Univ. Math. J. **32** (1983), 155–158.

[Kö]   G. Köthe, *Topological Vector Spaces I*, Springer-Verlag, 1969.

[KPS] S.G. Krein, Ju.I. Petunin, and E.M. Semenov, *Interpolation of Linear Operators*, Transl. Math. Monographs, vol. 54, 1982.

[La]   Q. Lai, *A note on weighted maximal inequalities*, Proc. Edimburgh Math. Soc. **40** (1997), 193–205.

[LN1] M.A. Leckband and C.J. Neugebauer, *A general maximal operator and the A_p-condition*, Trans. Amer. Math. Soc. **275** (1983), 821–831.

[LN2] M.A. Leckband and C.J. Neugebauer, *Weighted iterates and variants of the Hardy-Littlewood maximal operator*, Trans. Amer. Math. Soc. **279** (1983), 51–61.

[Lo1] G.G. Lorentz, *Some new functional spaces*, Ann. of Math. **51** (1950), 37–55.

[Lo2] G.G. Lorentz, *On the theory of spaces Λ*, Pacific J. Math. **1** (1951), 411–429.

[Mu]   B. Muckenhoupt, *Weighted norm inequalities for the Hardy maximal function*, Trans. Amer. Math. Soc. **165** (1972), 207–227.

[MW]   B. Muckenhoupt and R.L. Wheeden, *Weighted norm inequalities for fractional integrals*, Trans. Amer. Math. Soc. **192** (1974), 261–275.

[NO]   M. Nawrocki and A. Ortyński, *The Mackey topology and complemented subspaces of Lorentz sequence spaces $d(w,p)$ for $0 < p < 1$*, Trans. Amer. Math. Soc. **287** (1985), 713–722.

[Ne1] C.J. Neugebauer, *Weighted norm inequalities for average operators of monotone functions*, Publ. Mat. **35** (1991), 429–447.

[Ne2] C.J. Neugebauer, *Some classical operators on Lorentz spaces*, Forum Math. **4** (1992), 135–146.

[Po] N. Popa, *Basic sequences and subspaces in Lorentz sequence spaces without local convexity*, Trans. Amer. Math. Soc. **263** (1981), 431–456.

[Re] S. Reiner, *On the duals of Lorentz function and sequence spaces*, Indiana Univ. Math. J. **31** (1982), 65–72.

[Ru] W. Rudin, *Functional Analysis*, McGraw-Hill, Inc. International Series in Pure and Applied Mathematics, 1991.

[Sa] E. Sawyer, *Boundedness of classical operators on classical Lorentz spaces*, Studia Math. **96** (1990), 145–158.

[SS] G. Sinnamon and V. Stepanov, *The weighted Hardy inequality: new proofs and the case $p = 1$*, J. London Math. Soc. **154** (1996), 89–101.

[So] J. Soria, *Lorentz spaces of weak-type*, Quart. J. Math. Oxford **49** (1998), 93–103.

[Stn] E.M. Stein, *Harmonic Analysis. Real-Variable Methods, Orthogonality, and Oscillatory Integrals*, Princeton University Press, 1993.

[SW] E.M. Stein and G. Weiss, *Introduction to Fourier Analysis on Euclidean Spaces*, Princeton University Press, 1971.

[Stp1] V. Stepanov, *Integral operators on the cone of monotone functions*, Institute for Applied Mathematics. F-E.B. of the USSR Academy of Sciences, Khabarovsk, 1991.

[Stp2] V. Stepanov, *The weighted Hardy's inequality for nonincreasing functions*, Trans. Amer. Math. Soc. **338** (1993), 173–186.

Index

A, 5
A_d, 5
A_p, 108
$A_p(w)$, 91
A_p^*, 117
B_p, 16
$B_p(u)$, 91
$B_{p,\infty}$, 16
$B_{p,\infty}(u)$, 91
$B_{p_0,p_1,\infty}$, 16
B_{p_0,p_1}, 16
$L_0^\infty(X)$, 38
$L^{p,\infty}(X)$, 10
$L^{p,q}$, 25
$L^{p,q}(X)$, 10
$L^{p,q}(w)$, 26
M, 1, 5
Δ_2, 33
$\Delta_2(X)$, 33
$\Delta_2(\mu)$, 33
$\Gamma_X^p(w)$, 46
$\Gamma_X^{p,\infty}(d\Phi)$, 46
$\Gamma_X^{p,\infty}(w)$, 46
$\Lambda^p(w)$, 26
$\Lambda_X^{p,\infty}(w)$, 25
$\Lambda_X^p(w)$, 25
$\mathcal{S}(X)$, 40
$\mathcal{S}_0(X)$, 40
$\ell^p(w)$, 17
$\ell_{\text{dec}}^p(w)$, 18

$\ell^{p,\infty}(w)$, 18
λ_f, 6
$\Lambda_u^p(w)$, 90
$\Lambda_u^{p,\infty}(w)$, 91
$\Lambda_X^{p,q}(w)$, 27
L_{dec}^p, 6
L_{inc}^p, 6
$d(w,p)$, 25
$d^\infty(w,p)$, 26
f^*, 6
f^{**}, 6
$\mathcal{M}(X)$, 6
$\mathcal{M}^+(X)$, 6

absolutely continuous norm, 35
associate space, 43

Banach function space, 44

distribution function, 6, 26
dual space, 56

Fatou property, 7
functional lattice, 7

Hardy operator, 5
    discrete, 5, 17
Hardy-Littlewood maximal operator, 5, 91

Lorentz space, 10, 25
    Gamma, 46

weak-type, 25

Mackey topology, 57

nonincreasing rearrangement, 6, 25, 26

order continuous, 8

rearrangement invariant, 27, 44
regular class, 7
regular sequence, 49
regular weight, 49
resonant space, 34

step function, 69

weight, 6, 26
    doubling, 90, 102

Editorial Information

To be published in the *Memoirs*, a paper must be correct, new, nontrivial, and significant. Further, it must be well written and of interest to a substantial number of mathematicians. Piecemeal results, such as an inconclusive step toward an unproved major theorem or a minor variation on a known result, are in general not acceptable for publication.

Papers appearing in *Memoirs* are generally at least 80 and not more than 200 published pages in length. Papers less than 80 or more than 200 published pages require the approval of the Managing Editor of the Transactions/Memoirs Editorial Board.

As of February 28, 2007, the backlog for this journal was approximately 15 volumes. This estimate is the result of dividing the number of manuscripts for this journal in the Providence office that have not yet gone to the printer on the above date by the average number of monographs per volume over the previous twelve months, reduced by the number of volumes published in four months (the time necessary for preparing a volume for the printer). (There are 6 volumes per year, each usually containing at least 4 numbers.)

A Consent to Publish and Copyright Agreement is required before a paper will be published in the *Memoirs*. After a paper is accepted for publication, the Providence office will send a Consent to Publish and Copyright Agreement to all authors of the paper. By submitting a paper to the *Memoirs*, authors certify that the results have not been submitted to nor are they under consideration for publication by another journal, conference proceedings, or similar publication.

Information for Authors

Memoirs are printed from camera copy fully prepared by the author. This means that the finished book will look exactly like the copy submitted.

Initial submission. The AMS uses Centralized Manuscript Processing for initial submissions. Authors should submit a PDF file using the Initial Manuscript Submission form found at www.ams.org/cgi-bin/peertrack/submission.pl, or send one copy of the manuscript to the following address: Centralized Manuscript Processing, MEMOIRS OF THE AMS, 201 Charles Street, Providence, RI 02904-2294 USA. If a paper copy is being forwarded to the AMS, indicate that it is for it Memoirs and include the name of the corresponding author, contact information such as email address or mailing address, and the name of an appropriate Editor to review the paper (see the list of Editors below).

The paper must contain a *descriptive title* and an *abstract* that summarizes the article in language suitable for workers in the general field (algebra, analysis, etc.). The *descriptive title* should be short, but informative; useless or vague phrases such as "some remarks about" or "concerning" should be avoided. The *abstract* should be at least one complete sentence, and at most 300 words. Included with the footnotes to the paper should be the 2000 *Mathematics Subject Classification* representing the primary and secondary subjects of the article. The classifications are accessible from www.ams.org/msc/. The list of classifications is also available in print starting with the 1999 annual index of *Mathematical Reviews*. The Mathematics Subject Classification footnote may be followed by a list of *key words and phrases* describing the subject matter of the article and taken from it. Journal abbreviations used in bibliographies are listed in the latest *Mathematical Reviews* annual index. The series abbreviations are also accessible from www.ams.org/publications/. To help in preparing and verifying references, the AMS offers MR Lookup, a Reference Tool for Linking, at www.ams.org/mrlookup/.

Electronically prepared manuscripts. The AMS encourages electronically prepared manuscripts, with a strong preference for $\mathcal{AMS}$-LaTeX. To this end, the Society has prepared $\mathcal{AMS}$-LaTeX author packages for each AMS publication. Author packages include instructions for preparing electronic manuscripts, samples, and a style file that generates the particular design specifications of that publication series. Though $\mathcal{AMS}$-LaTeX is the highly preferred format of TeX, author packages are also available in $\mathcal{AMS}$-TeX.

Authors may retrieve an author package from the AMS website starting from www.ams.org/tex/ or via FTP to ftp.ams.org (login as anonymous, enter username as password, and type cd pub/author-info). The *AMS Author Handbook* and the *Instruction Manual* are available in PDF format following the author packages link from www.ams.org/tex/. The author package can also be obtained free of charge by sending email to tech-support@ams.org (Internet) or from the Publication Division, American Mathematical Society, 201 Charles St., Providence, RI 02904-2294, USA. When requesting an author package, please specify $\mathcal{AMS}$-LaTeX or $\mathcal{AMS}$-TeX and the publication in which your paper will appear. Please be sure to include your complete mailing address.

After acceptance. The final version of the electronic file should be sent to the Providence office (this includes any TeX source file, any graphics files, and the DVI or PostScript file) immediately after the paper has been accepted for publication.

Before sending the source file, be sure you have proofread your paper carefully. The files you send must be the EXACT files used to generate the proof copy that was accepted for publication. For all publications, authors are required to send a printed copy of their paper, which exactly matches the copy approved for publication, along with any graphics that will appear in the paper.

Accepted electronically prepared files can be submitted via the web at www.ams.org/submit-book-journal/, sent via FTP, or sent on CD-Rom or diskette to the Electronic Prepress Department, American Mathematical Society, 201 Charles Street, Providence, RI 02904-2294 USA. TeX source files, DVI files, and PostScript files can be transferred over the Internet by FTP to the Internet node ftp.ams.org (130.44.1.100). When sending a manuscript electronically via CD-Rom or diskette, please be sure to include a message identifying the paper as a Memoir.

Electronically prepared manuscripts can also be sent via email to pub-submit@ams.org (Internet). In order to send files via email, they must be encoded properly. (DVI files are binary and PostScript files tend to be very large.)

Electronic graphics. Comprehensive instructions on preparing graphics are available at www.ams.org/jourhtml/. A few of the major requirements are given here.

Submit files for graphics as EPS (Encapsulated PostScript) files. This includes graphics originated via a graphics application as well as scanned photographs or other computer-generated images. If this is not possible, TIFF files are acceptable as long as they can be opened in Adobe Photoshop or Illustrator. No matter what method was used to produce the graphic, it is necessary to provide a paper copy to the AMS.

Authors using graphics packages for the creation of electronic art should also avoid the use of any lines thinner than 0.5 points in width. Many graphics packages allow the user to specify a "hairline" for a very thin line. Hairlines often look acceptable when proofed on a typical laser printer. However, when produced on a high-resolution laser imagesetter, hairlines become nearly invisible and will be lost entirely in the final printing process.

Screens should be set to values between 15% and 85%. Screens which fall outside of this range are too light or too dark to print correctly. Variations of screens within a graphic should be no less than 10%.

Inquiries. Any inquiries concerning a paper that has been accepted for publication should be sent to memo-query@ams.org or directly to the Electronic Prepress Department, American Mathematical Society, 201 Charles St., Providence, RI 02904-2294 USA.

Editors

This journal is designed particularly for long research papers, normally at least 80 pages in length, and groups of cognate papers in pure and applied mathematics. Papers intended for publication in the *Memoirs* should be addressed to one of the following editors. The AMS uses Centralized Manuscript Processing for initial submissions to AMS journals. Authors should follow instructions listed on the Initial Submission page found at www.ams.org/memo/memosubmit.html.

Algebra to ALEXANDER KLESHCHEV, Department of Mathematics, University of Oregon, Eugene, OR 97403-1222; email: ams@noether.uoregon.edu

Algebra and its application to MINA TEICHER, Emmy Noether Research Institute for Mathematics, Bar-Ilan University, Ramat-Gan 52900, Israel; email: teicher@macs.biu.ac.il

Algebraic geometry to DAN ABRAMOVICH, Department of Mathematics, Brown University, Box 1917, Providence, RI 02912; email: amsedit@math.brown.edu

Algebraic number theory to V. KUMAR MURTY, Department of Mathematics, University of Toronto, 100 St. George Street, Toronto, ON M5S 1A1, Canada; email: murty@math.toronto.edu

Algebraic topology to ALEJANDRO ADEM, Department of Mathematics, University of British Columbia, Room 121, 1984 Mathematics Road, Vancouver, British Columbia, Canada V6T 1Z2; email: adem@math.ubc.ca

Combinatorics to JOHN R. STEMBRIDGE, Department of Mathematics, University of Michigan, Ann Arbor, Michigan 48109-1109; email: FRS@umich.edu

Complex analysis and harmonic analysis to ALEXANDER NAGEL, Department of Mathematics, University of Wisconsin, 480 Lincoln Drive, Madison, WI 53706-1313; email: nagel@math.wisc.edu

Differential geometry and global analysis to LISA C. JEFFREY, Department of Mathematics, University of Toronto, 100 St. George St., Toronto, ON Canada M5S 3G3; email: jeffrey@math.toronto.edu

Dynamical systems and ergodic theory to AMIE WILKINSON, Department of Mathematics, Northwestern University, 2033 Sheridan Road, Evanston, IL 60208-2730; email: transactions@math.northwestern.edu

Functional analysis and operator algebras to DIMITRI SHLYAKHTENKO, Department of Mathematics, University of California, Los Angeles, CA 90095; email: shlyakht@math.ucla.edu

Geometric analysis to WILLIAM P. MINICOZZI II, Department of Mathematics, Johns Hopkins University, 3400 N. Charles St., Baltimore, MD 21218; email: trans@math.jhu.edu

Geometric analysis to MLADEN BESTVINA, Department of Mathematics, University of Utah, 155 South 1400 East, JWB 233, Salt Lake City, Utah 84112-0090; email: bestvina@math.utah.edu

Harmonic analysis, representation theory, and Lie theory to ROBERT J. STANTON, Department of Mathematics, The Ohio State University, 231 West 18th Avenue, Columbus, OH 43210-1174; email: stanton@math.ohio-state.edu

Logic to STEFFEN LEMPP, Department of Mathematics, University of Wisconsin, 480 Lincoln Drive, Madison, Wisconsin 53706-1388; email: lempp@math.wisc.edu

Partial differential equations to GUSTAVO PONCE, Department of Mathematics, South Hall, Room 6607, University of California, Santa Barbara, CA 93106; email: ponce@math.ucsb.edu

Partial differential equations and dynamical systems to PETER POLACIK, School of Mathematics, University of Minnesota, Minneapolis, MN 55455; email: polacik@math.umn.edu

Probability and statistics to KRZYSZTOF BURDZY, Department of Mathematics, University of Washington, Box 354350, Seattle, Washington 98195-4350; email: burdzy@math.washington.edu

Real analysis and partial differential equations to DANIEL TATARU, Department of Mathematics, University of California, Berkeley, Berkeley, CA 94720; email: tataru@math.berkeley.edu

All other communications to the editors should be addressed to the Managing Editor, ROBERT GURALNICK, Department of Mathematics, University of Southern California, Los Angeles, CA 90089-1113; email: guralnic@math.usc.edu.

Titles in This Series

879 **O. García-Prada, P. B. Gothen, and V. Muñoz,** Betti numbers of the moduli space of rank 3 parabolic Higgs bundles, 2007

878 **Alessandra Celletti and Luigi Chierchia,** KAM stability and celestial mechanics, 2007

877 **María J. Carro, José A. Raposo, and Javier Soria,** Recent developments in the theory of Lorentz spaces and weighted inequalities, 2007

876 **Gabriel Debs and Jean Saint Raymond,** Borel liftings of Borel sets: Some decidable and undecidable statements, 2007

875 **C. Krattenthaler and T. Rivoal,** Hypergéométrie et fonction zêta de Riemann, 2007

874 **Sonia Natale,** Semisolvability of semisimple Hopf algebras of low dimension, 2007

873 **A. J. Duncan,** Exponential genus problems in one-relator products of groups, 2007

872 **Anthony V. Geramita, Tadahito Harima, Juan C. Migliore, and Yong Su Shin,** The Hilbert function of a level algebra, 2007

871 **Pascal Auscher,** On necessary and sufficient conditions for L^p-estimates of Riesz transforms associated to elliptic operators on $\mathbb{R}^n$ and related estimates, 2007

870 **Takuro Mochizuki,** Asymptotic behaviour of tame harmonic bundles and an application to pure twistor D-modules, Part 2, 2007

869 **Takuro Mochizuki,** Asymptotic behaviour of tame harmonic bundles and an application to pure twistor D-modules, Part 1, 2007

868 **Gelu Popescu,** Entropy and multivariable interpolation, 2006

867 **Vilmos Totik,** Metric properties of harmonic measures, 2006

866 **William Craig,** Semigroups underlying first-order logic, 2006

865 **Nathanial P. Brown,** Invariant means and finite representation theory of $C*$-algebras, 2006

864 **John M. Lee,** Fredholm operators and Einstein metrics on conformally compact manifolds, 2006

863 **M. Lübke and A. Teleman,** The Universal Kobayashi-Hitchin correspondence on Hermitian manifolds, 2006

862 **Alberto Canonaco,** The Beilinson complex and canonical rings of irregular surfaces, 2006

861 **Leon A. Takhtajan and Lee-Peng Teo,** Weil-Petersson metric on the universal Teichmüller space, 2006

860 **Thomas M. Fiore,** Pseudo limits, biadjoints and pseudo algebras: Categorical foundations of conformal field theory, 2006

859 **N. Arcozzi, R. Rochberg, and E. Sawyer,** Carleson measures and interpolating sequences for Besov spaces on complex balls, 2006

858 **Enrico Valdinoci, Berardino Sciunzi, and Vasile Ovidiu Savin,** Flat level set regularity of p-Laplace phase transitions, 2006

857 **Donatella Danielli, Nocola Garofalo, and Duy-Minh Nhieu,** Non-doubling Ahlfors measures, perimeter measures, and the characterization of the trace spaces of Sobolev functions in Carnot-Carathéodory spaces, 2006

856 **Vladimir Bolotnikov and Harry Dym,** On boundary interpolation for matrix valued Schur functions, 2006

855 **Yevgenia Kashina, Yorck Sommerhäuser, and Yongchang Zhu,** On higher Frobenius-Schur indicators, 2006

For a complete list of titles in this series, visit the
AMS Bookstore at **www.ams.org/bookstore/**.